double happiness

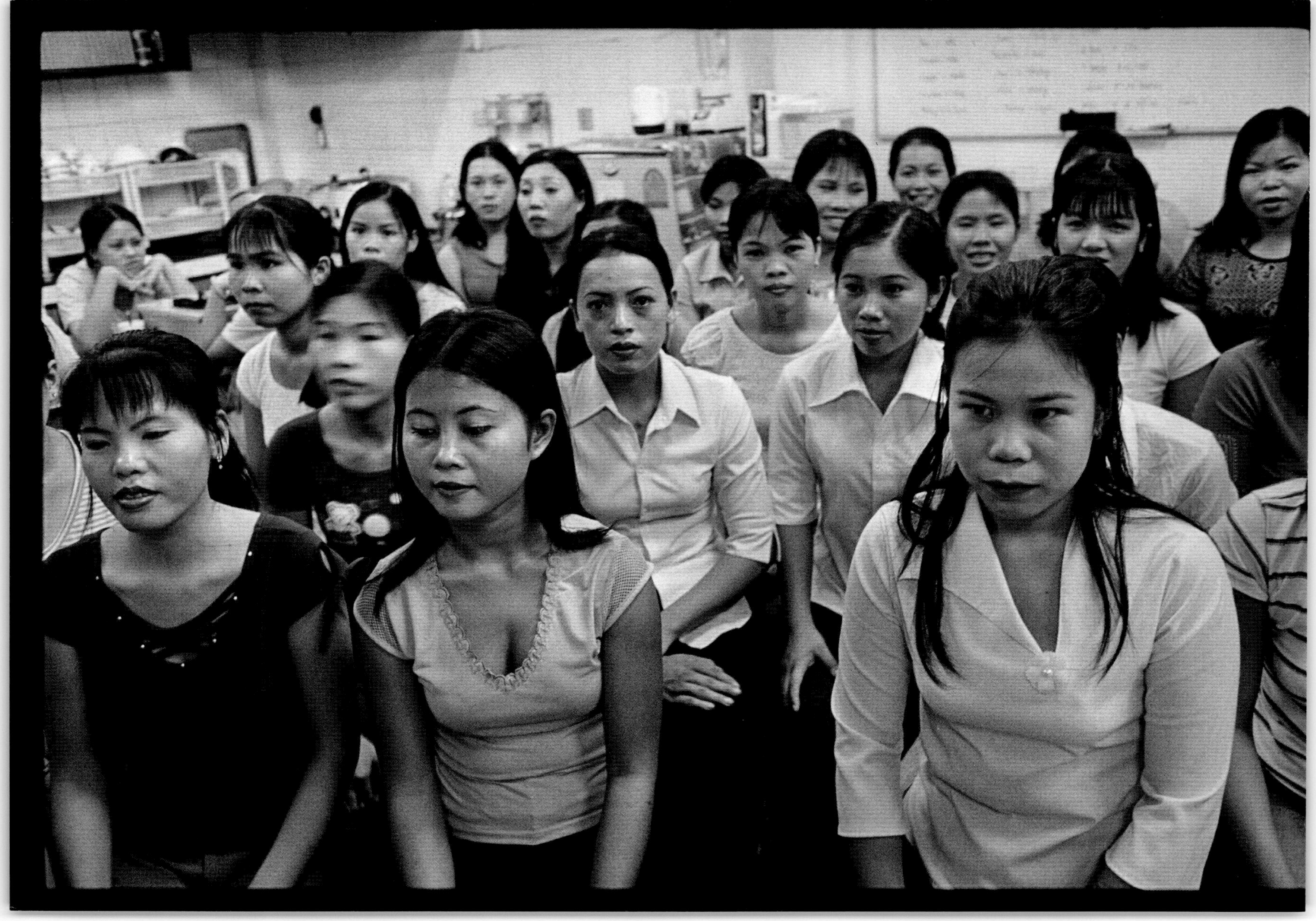

double happiness

photographs by chien-chi chang

essay by claudia glenn dowling

aperture

Marriage brokers recruit young Vietnamese women to come to Ho Chi Minh City where they are viewed by groups of Taiwanese men. Each of the men pays a fee to pick a suitable bride from the line-up. Within days of meeting, the couples will be married.

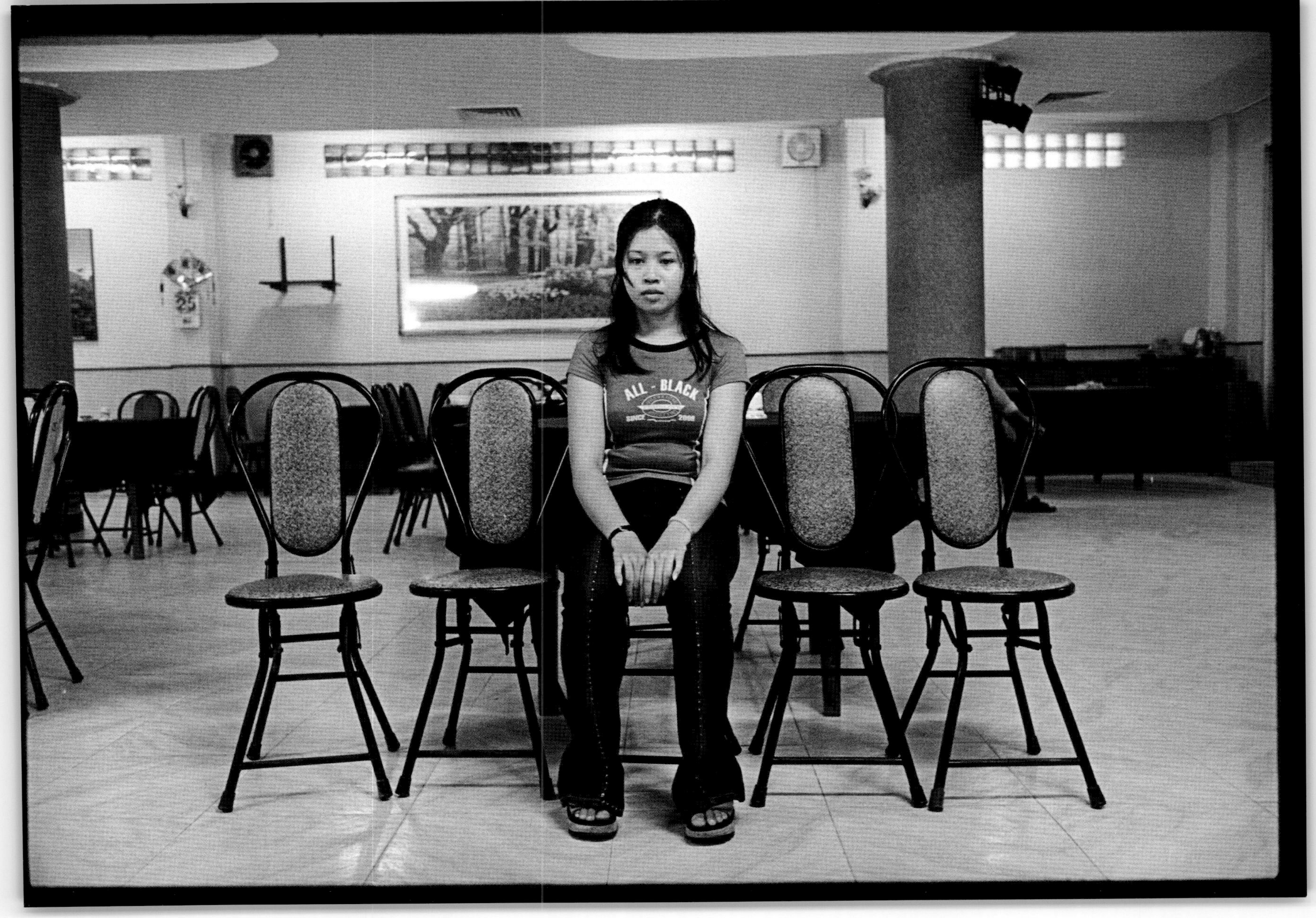
ALL - BLACK

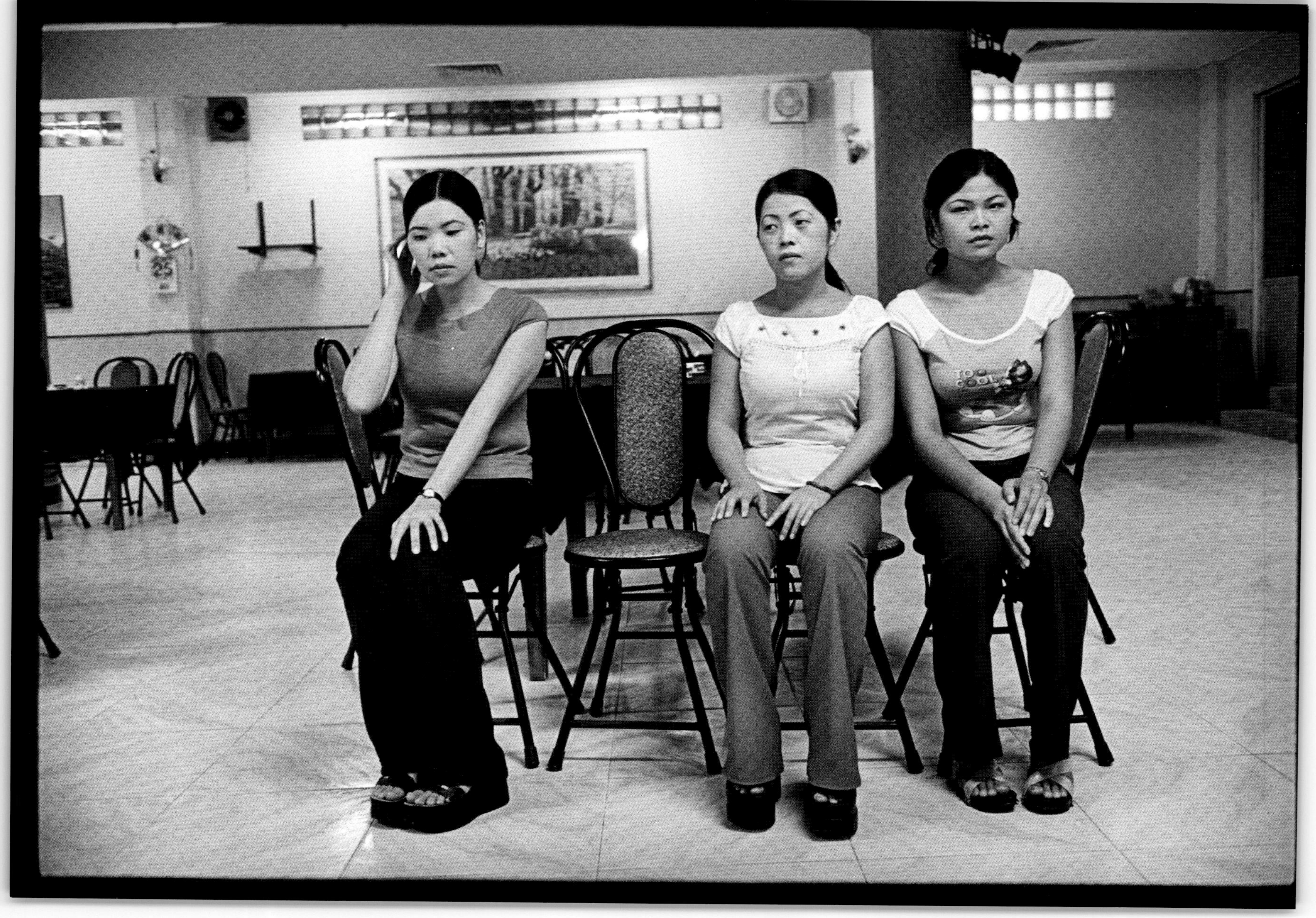

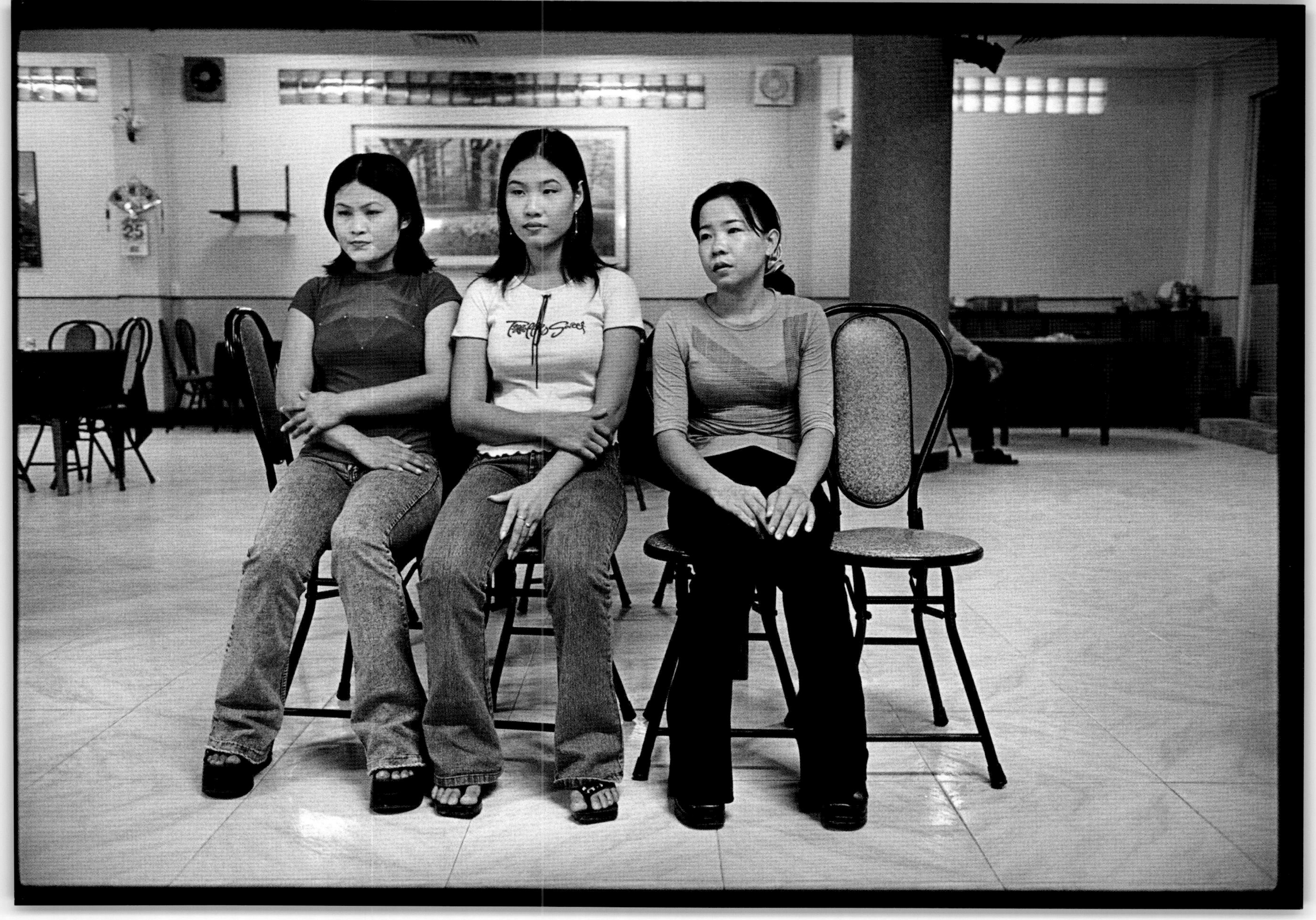

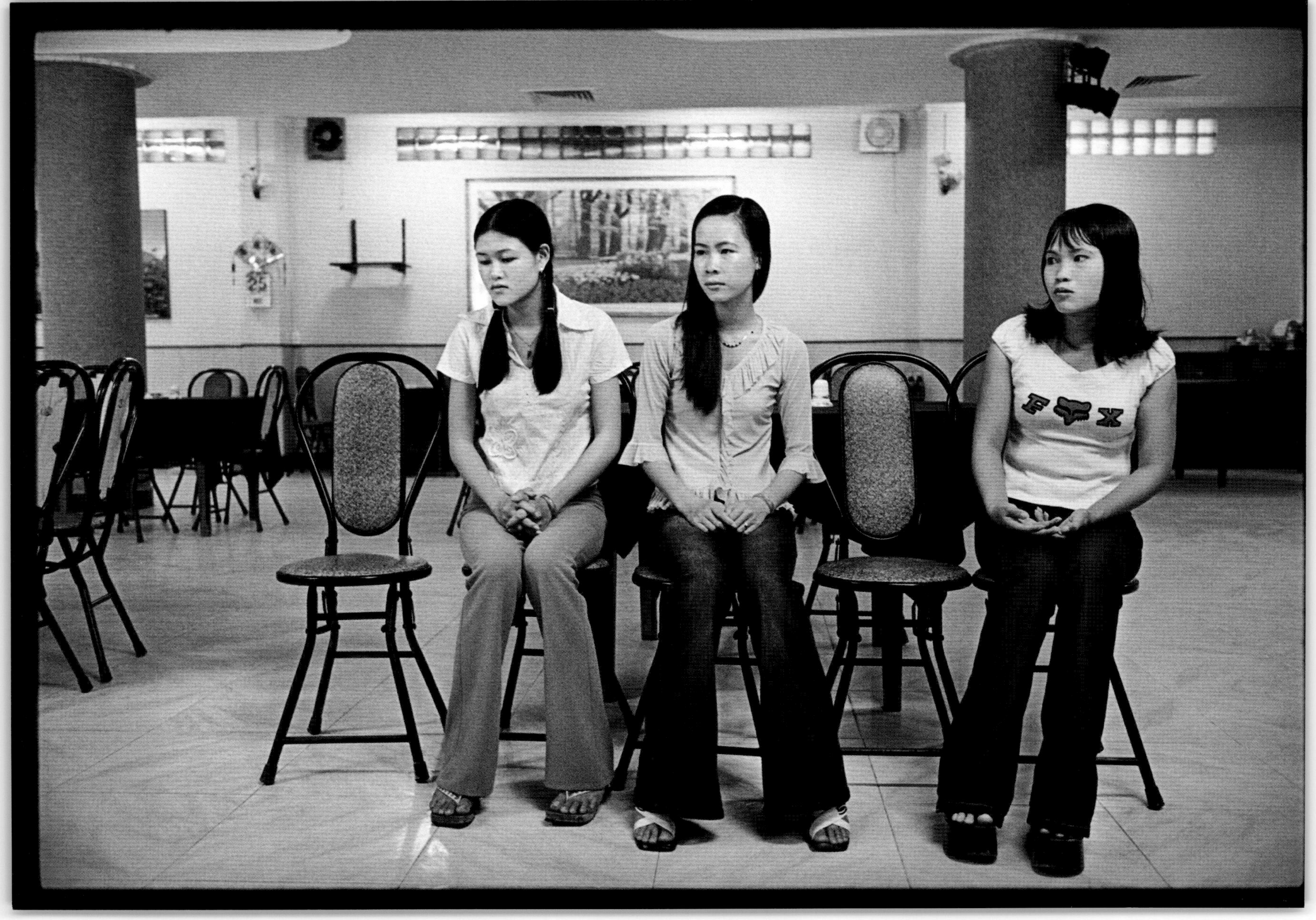

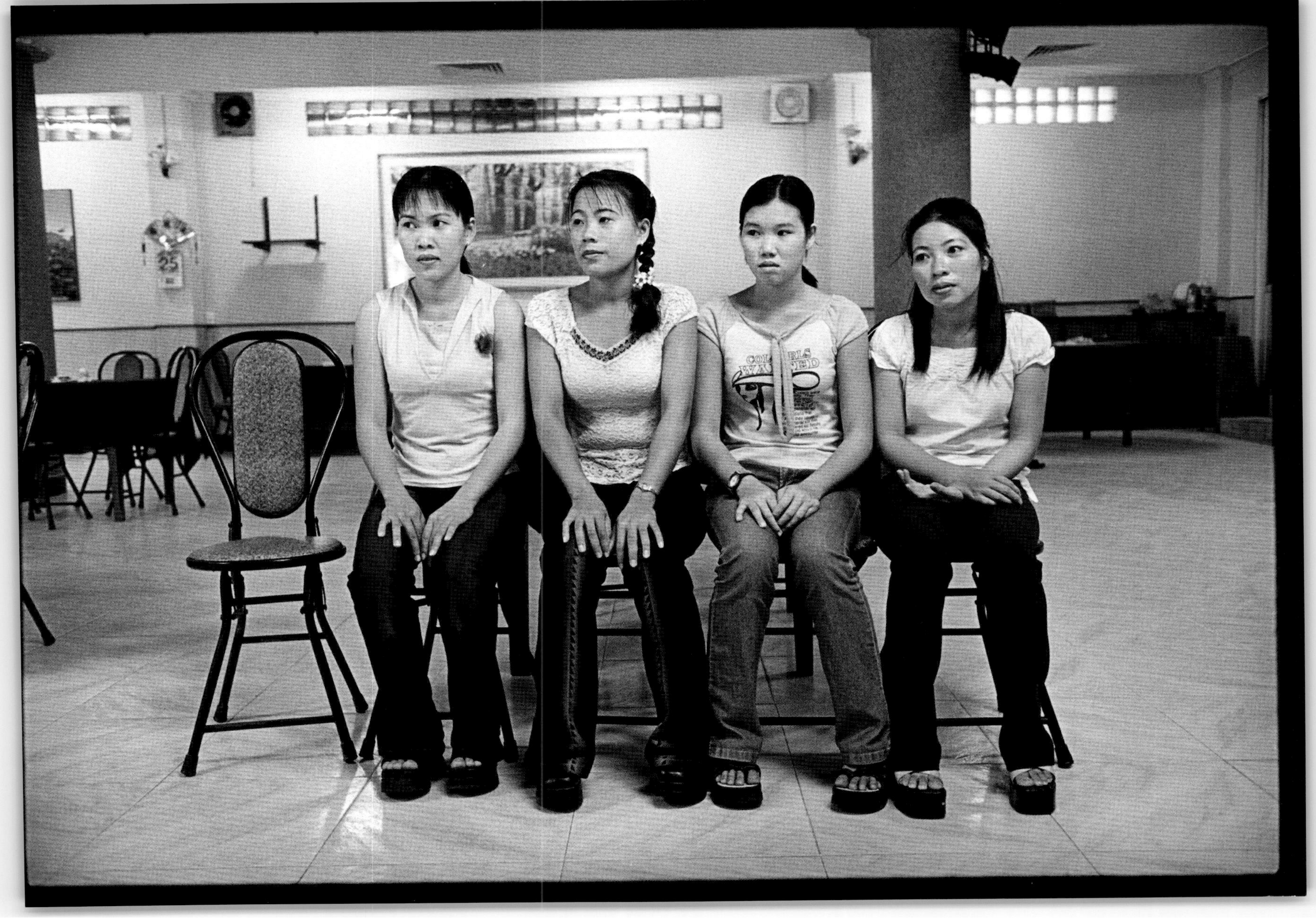

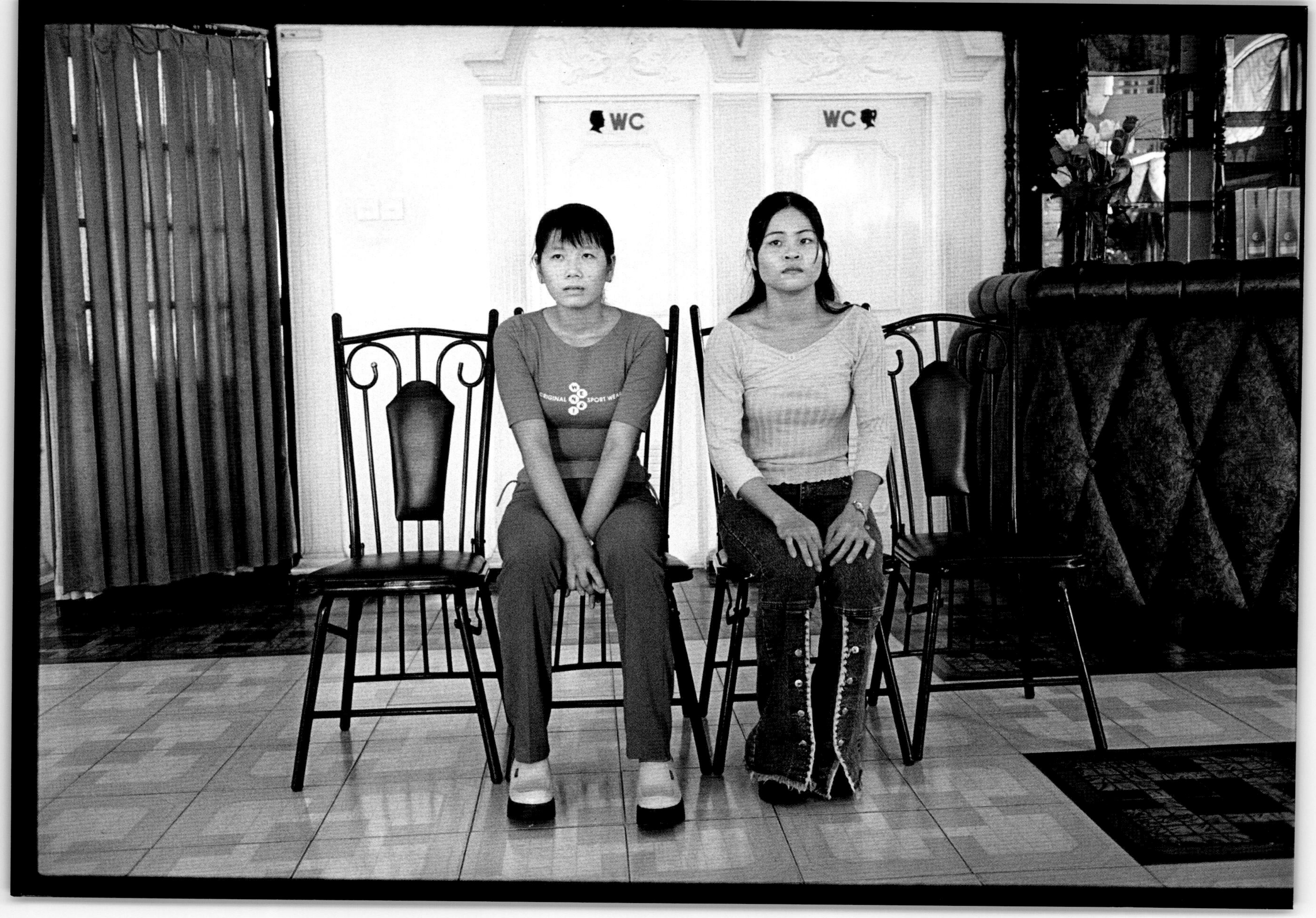
WC
WC

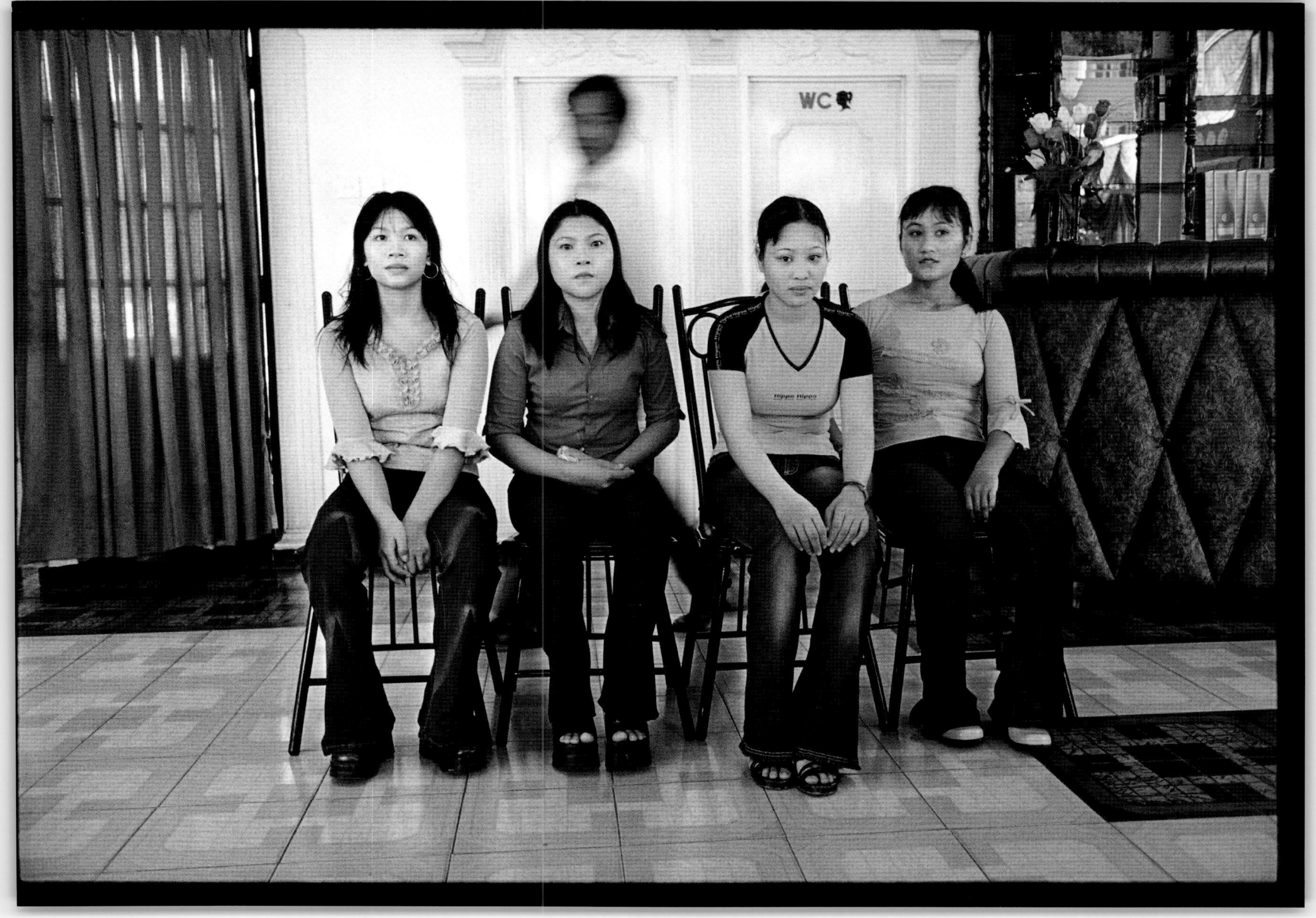
WC

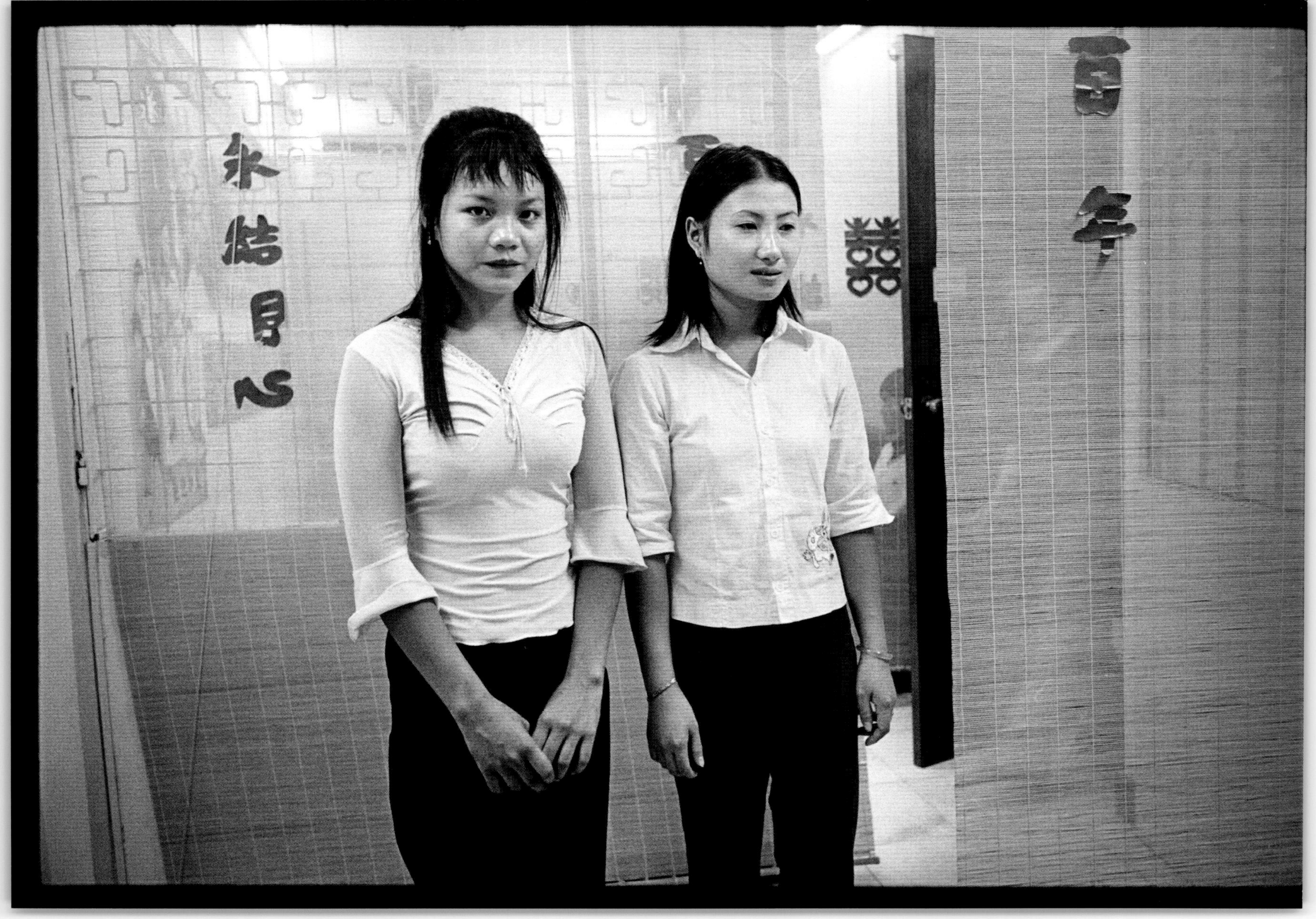
永结同心
囍
百年

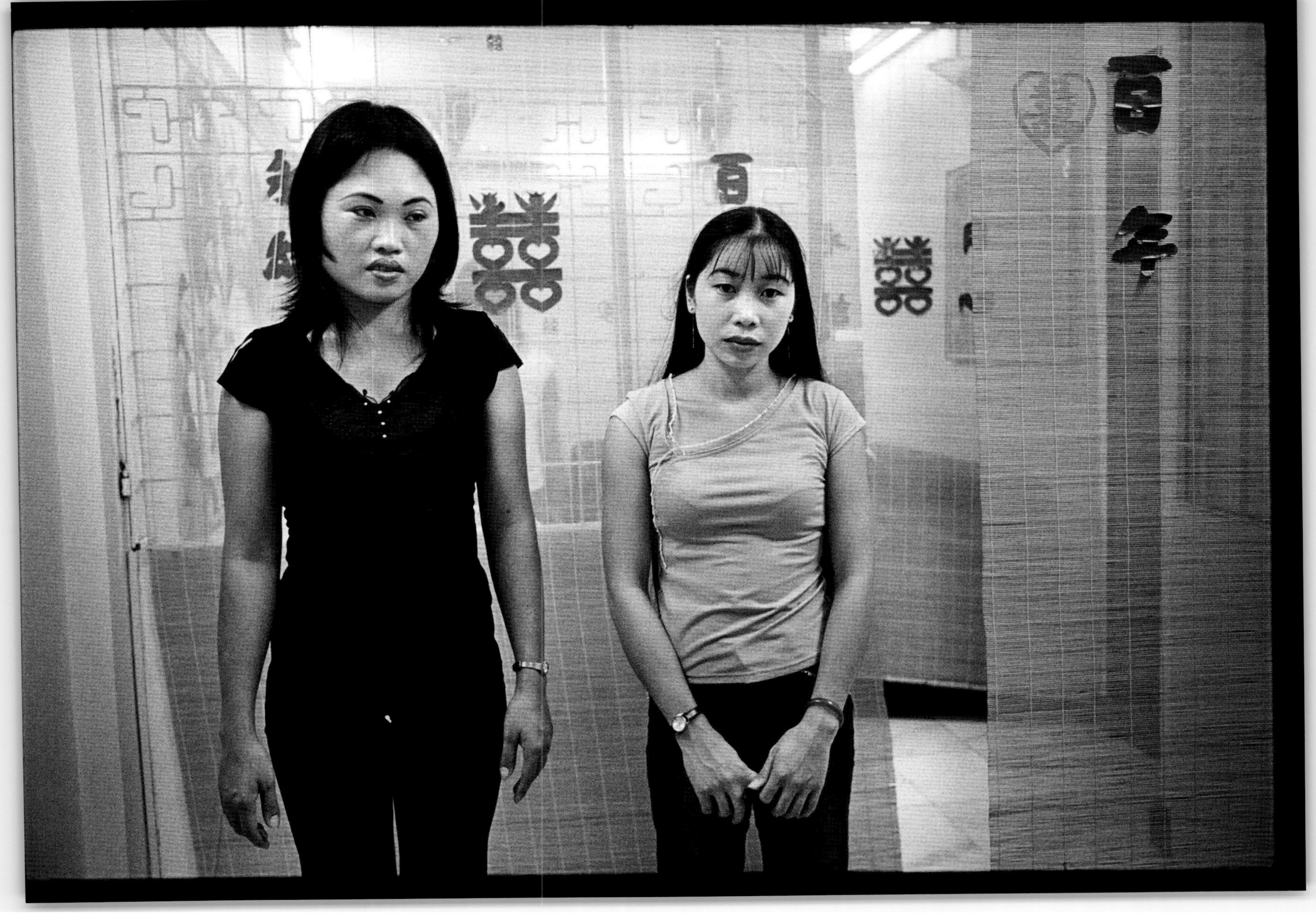
囍
百
囍
百年

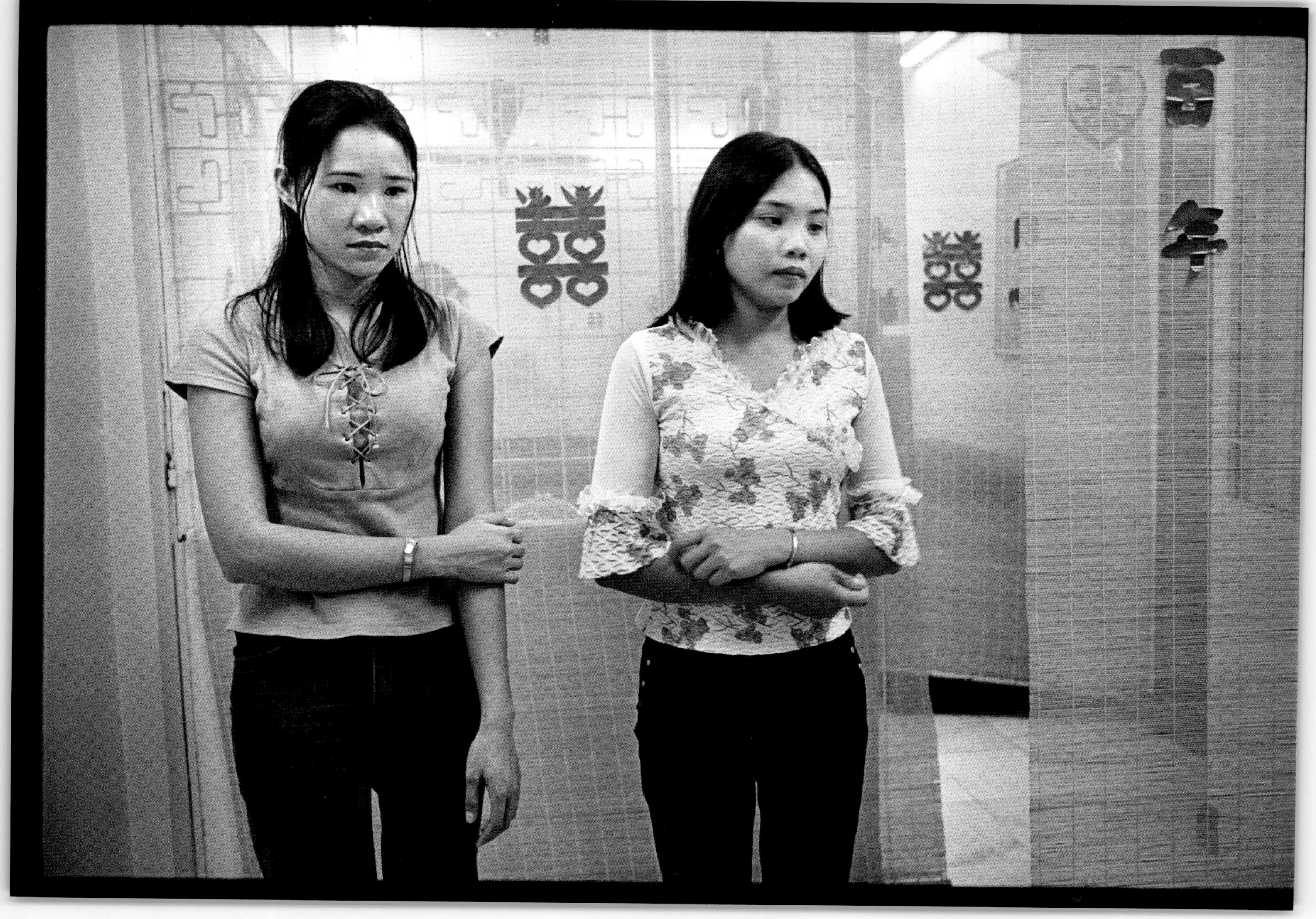
囍
囍
百年

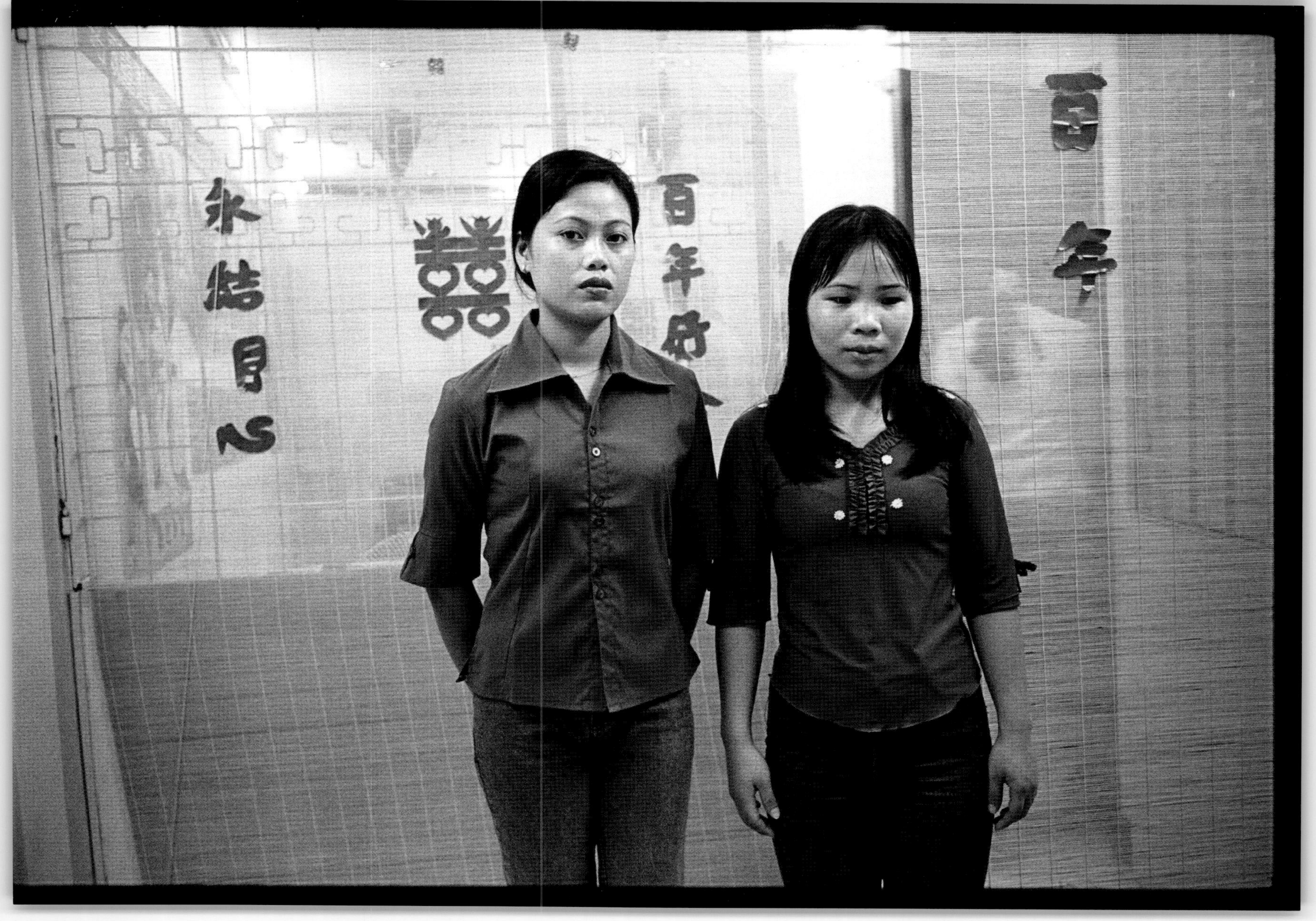

囍
百年

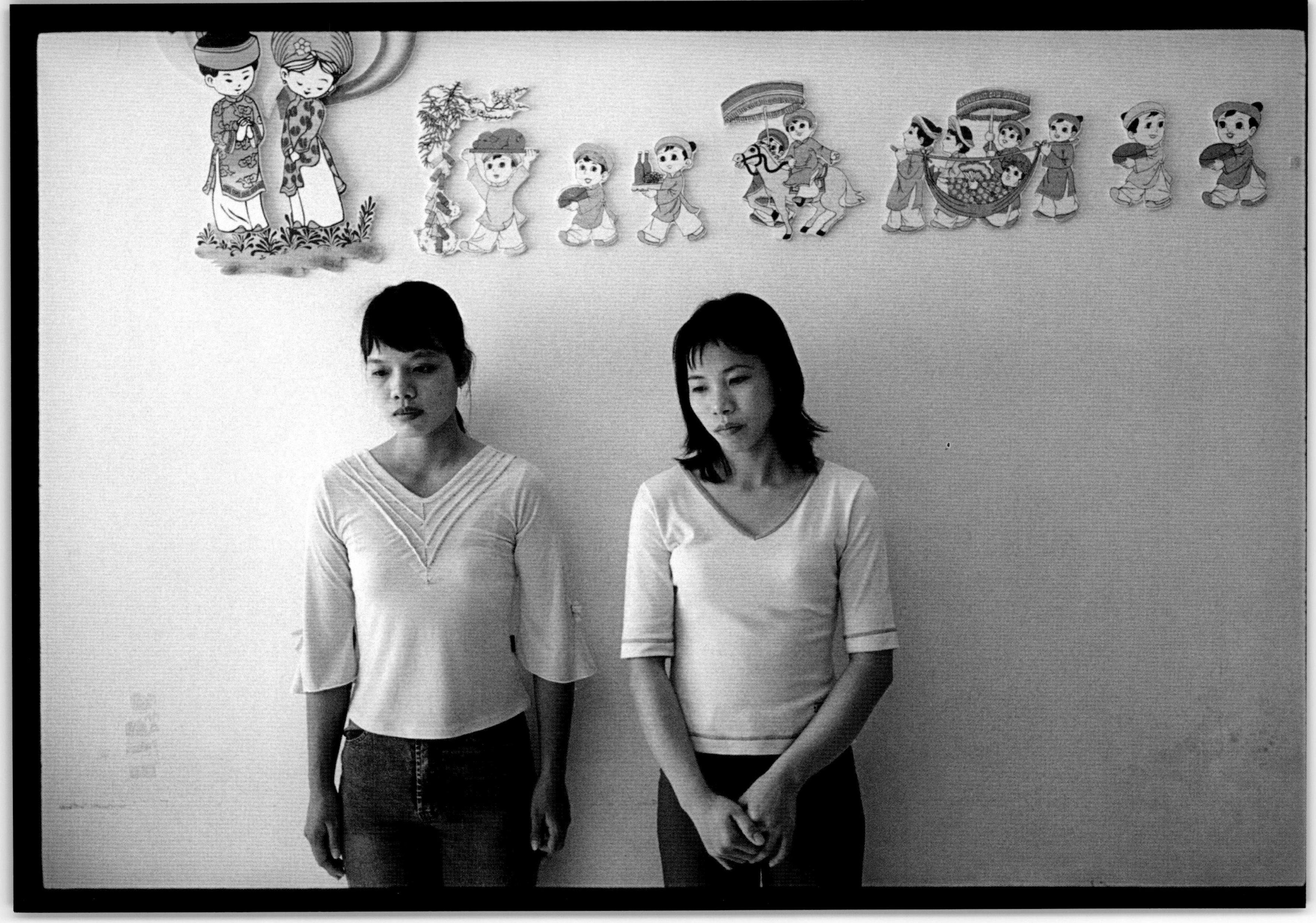

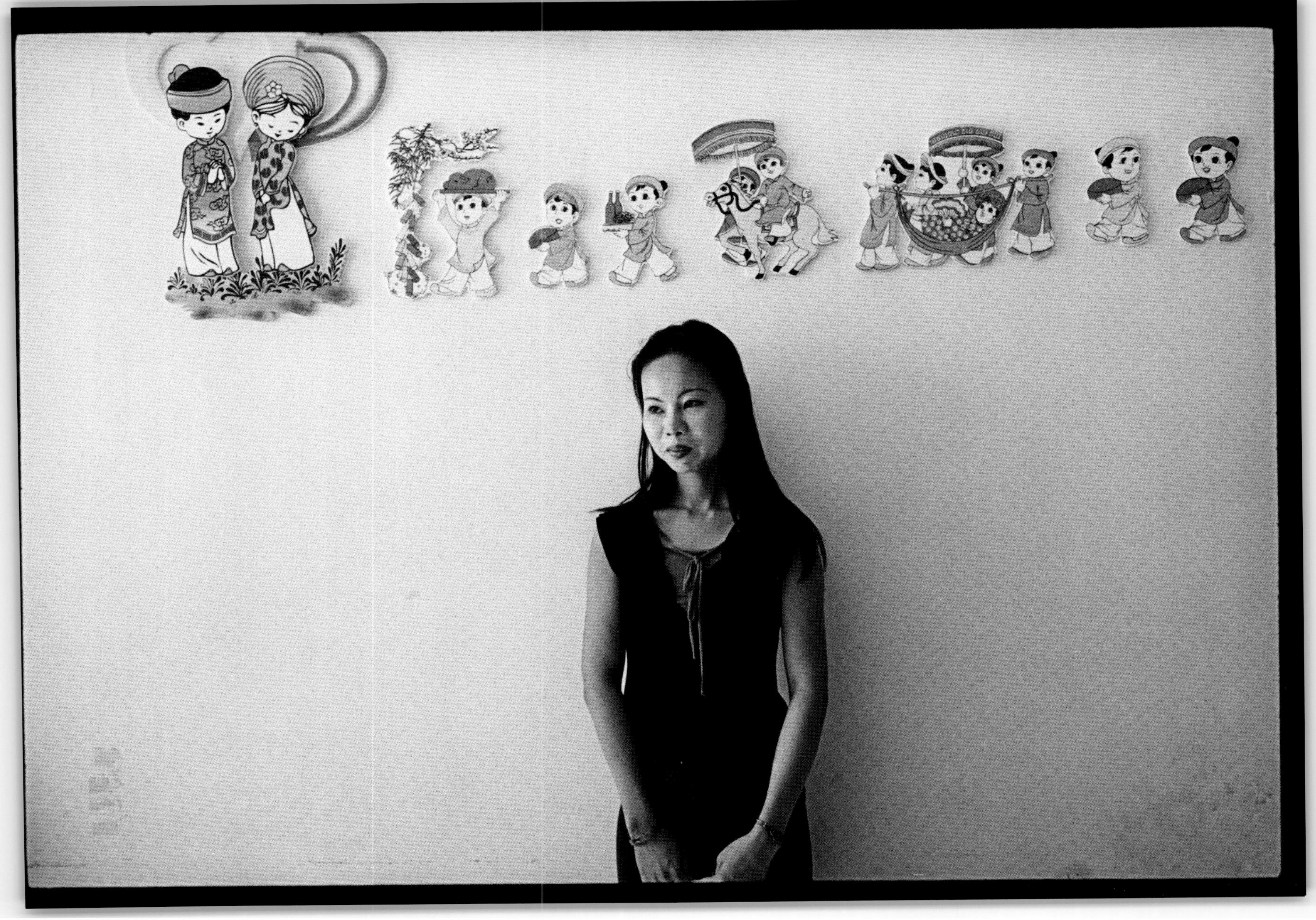

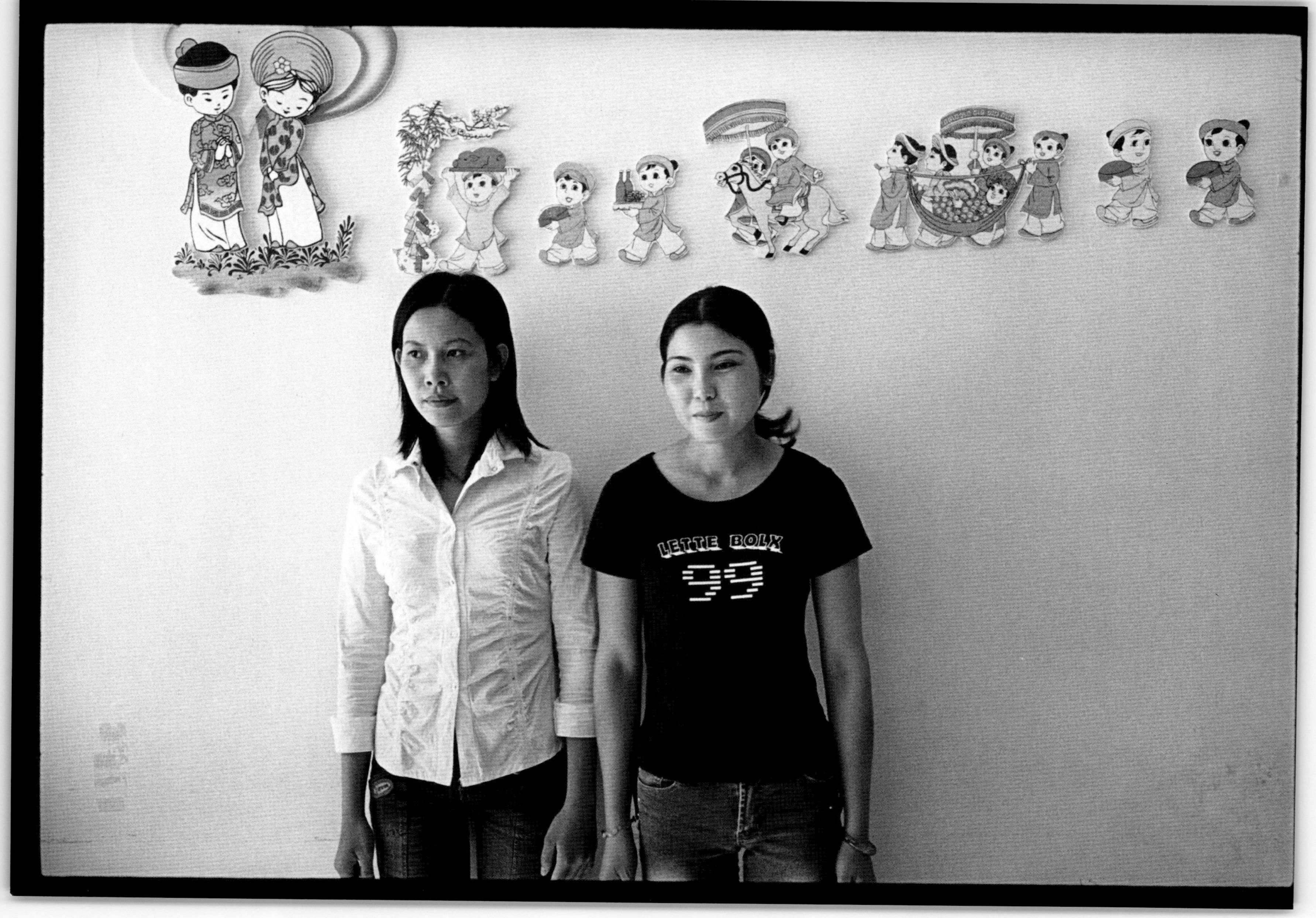
LETTE BOLX
99

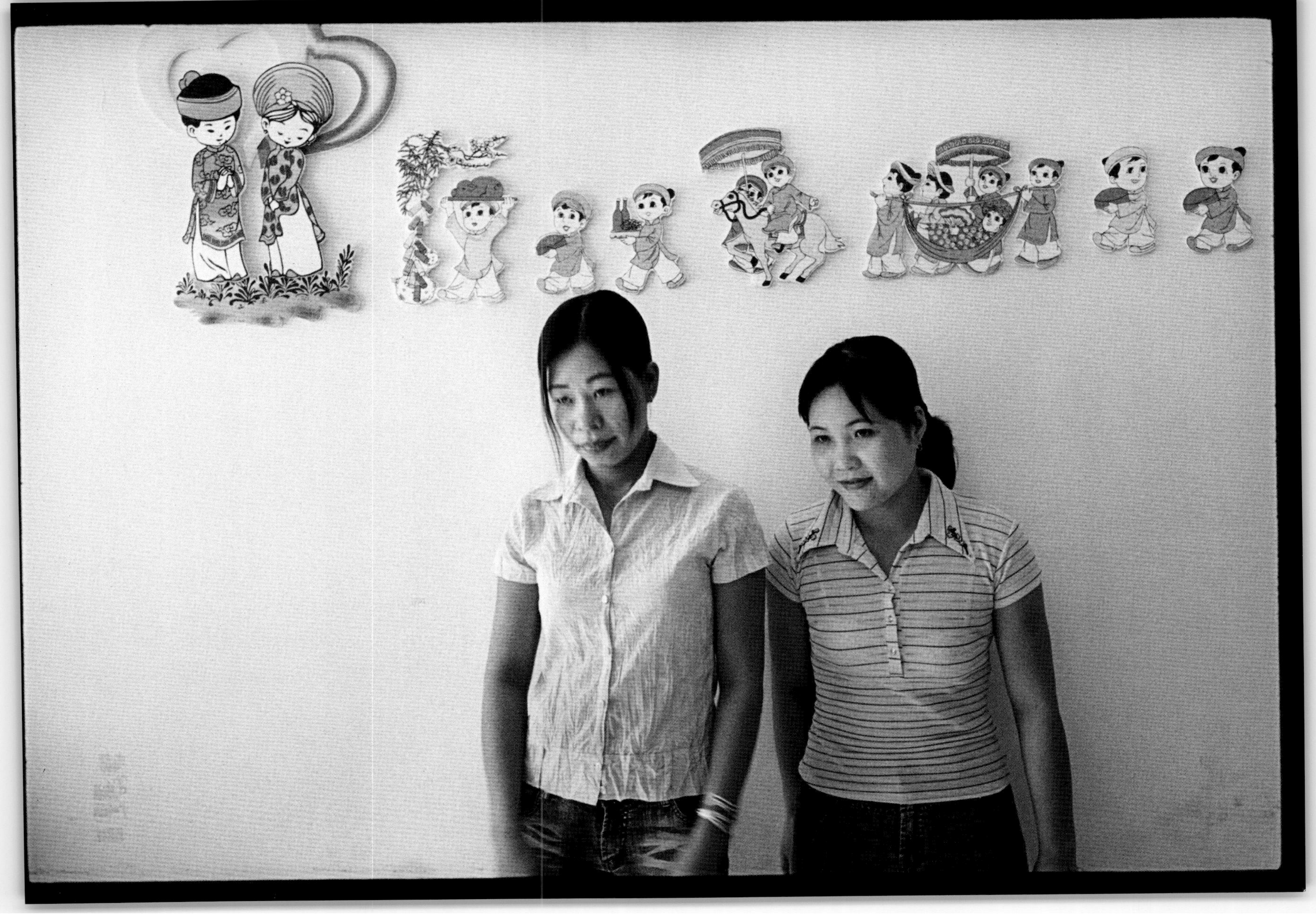

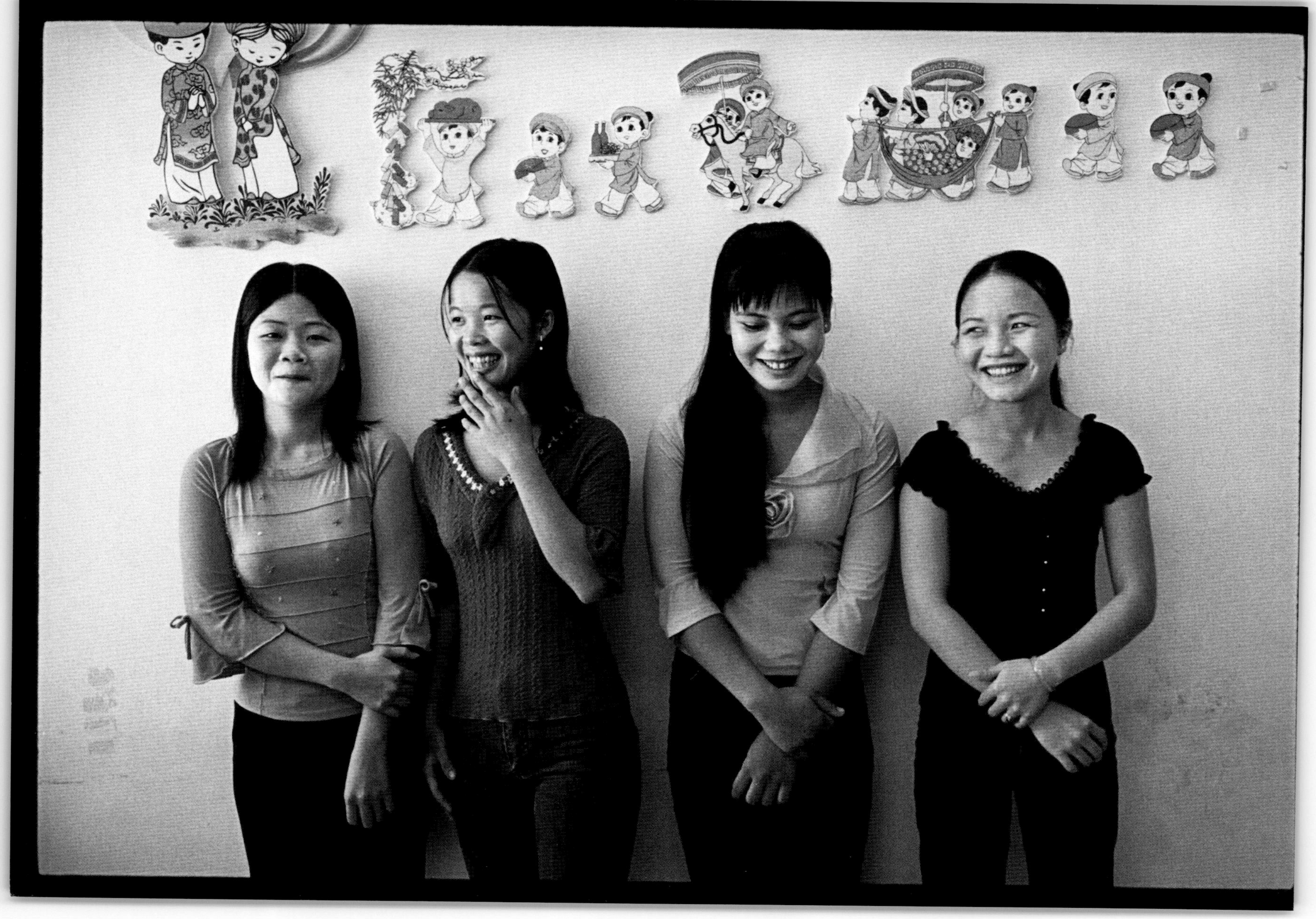

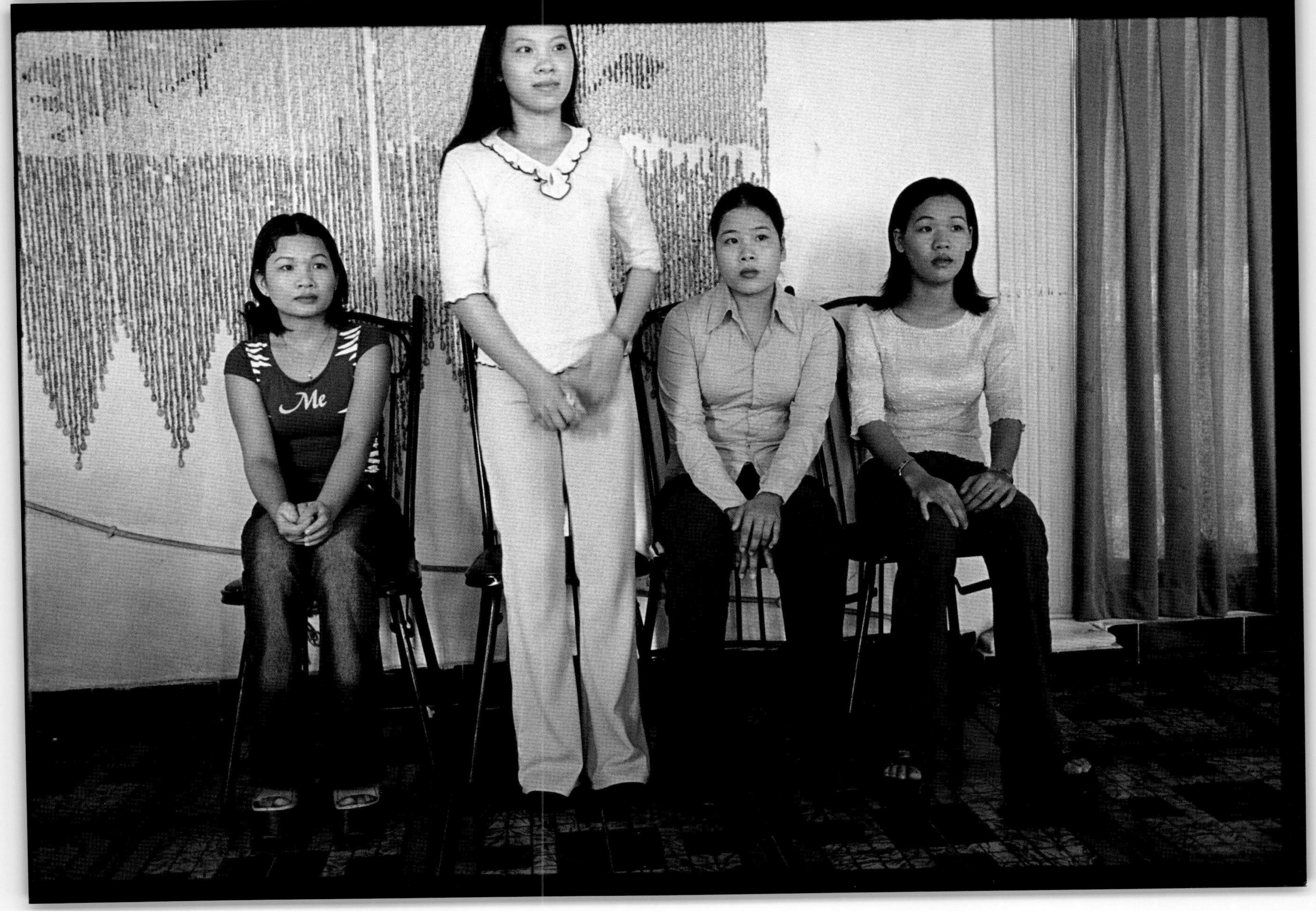
Me

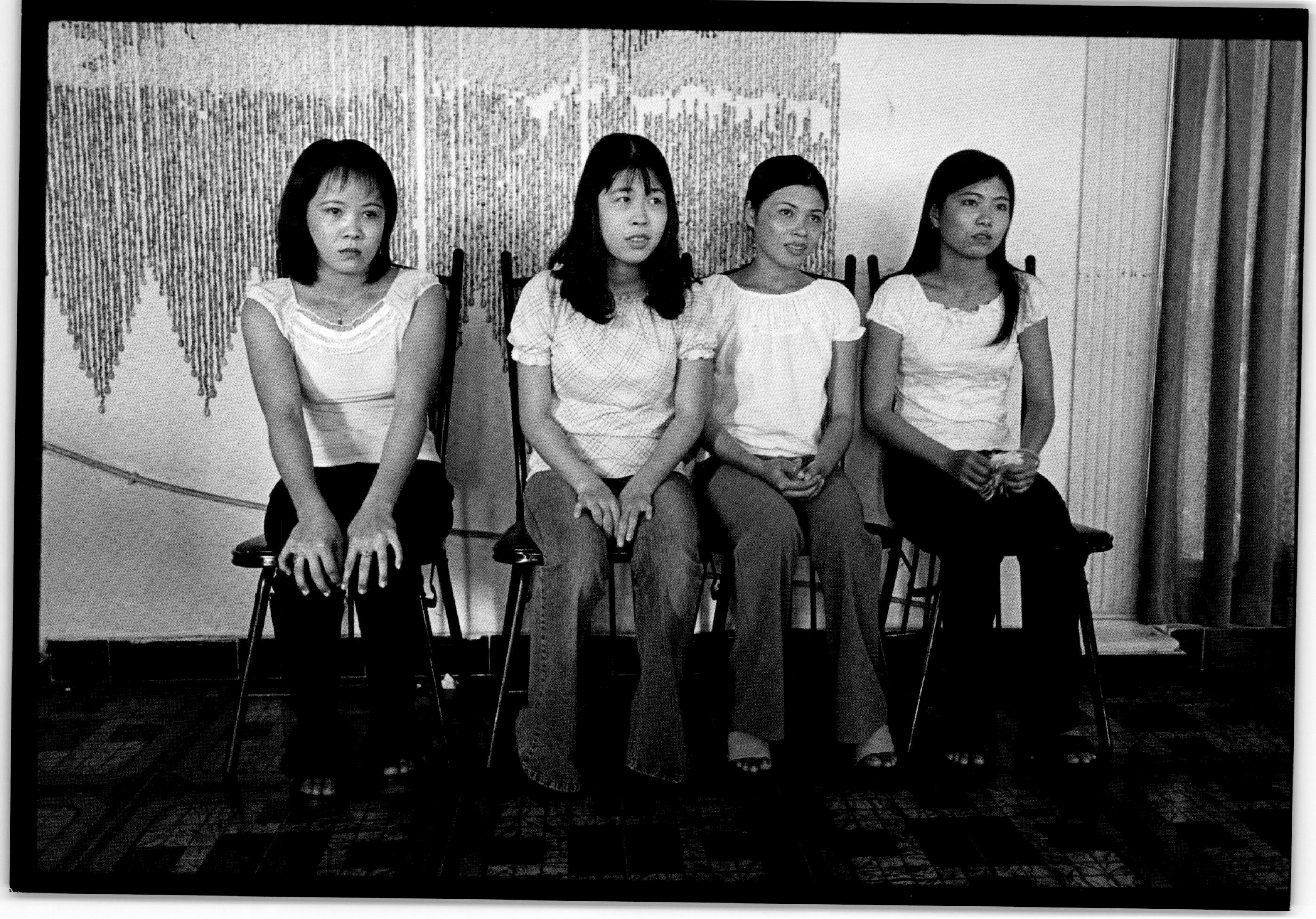

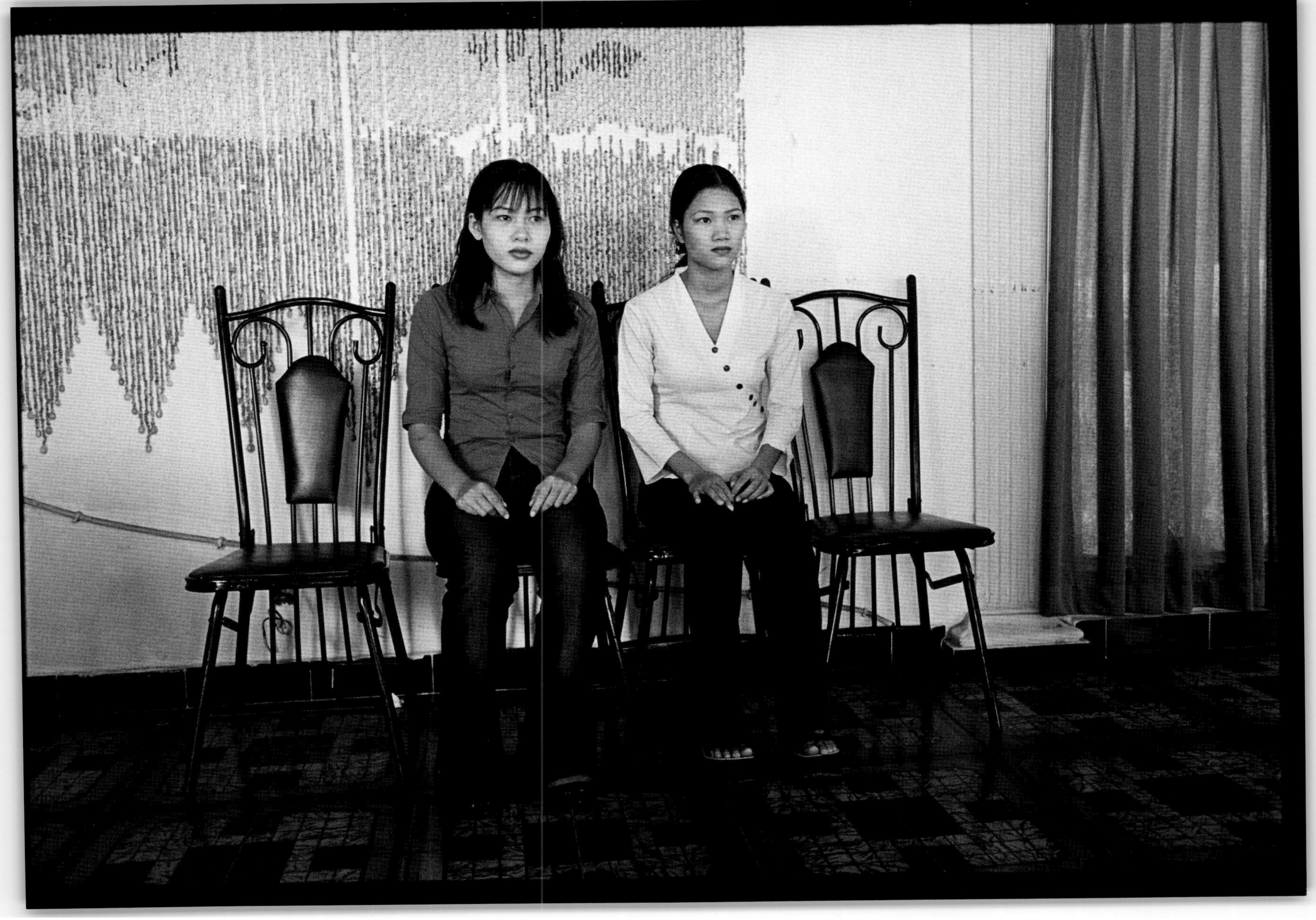

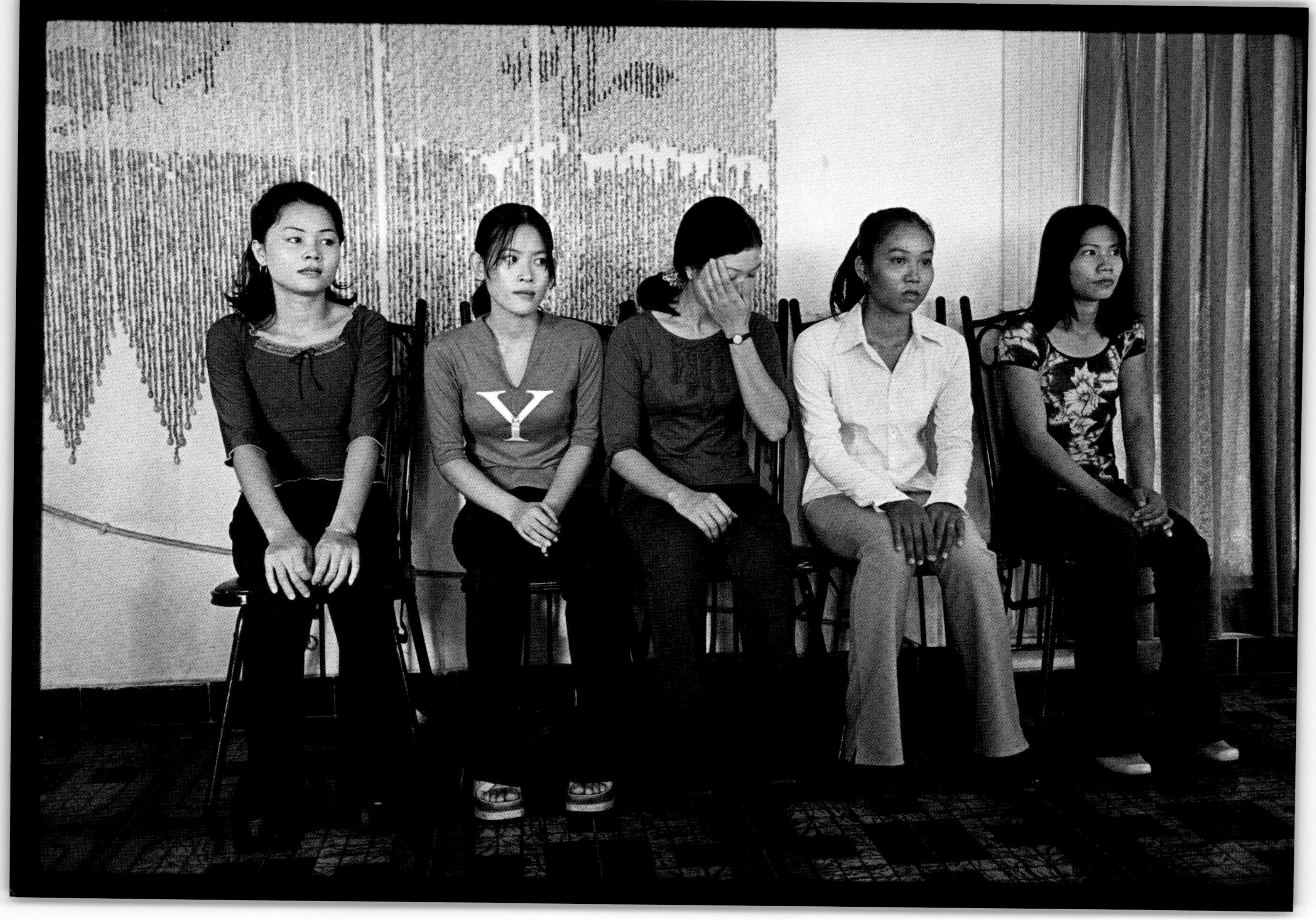

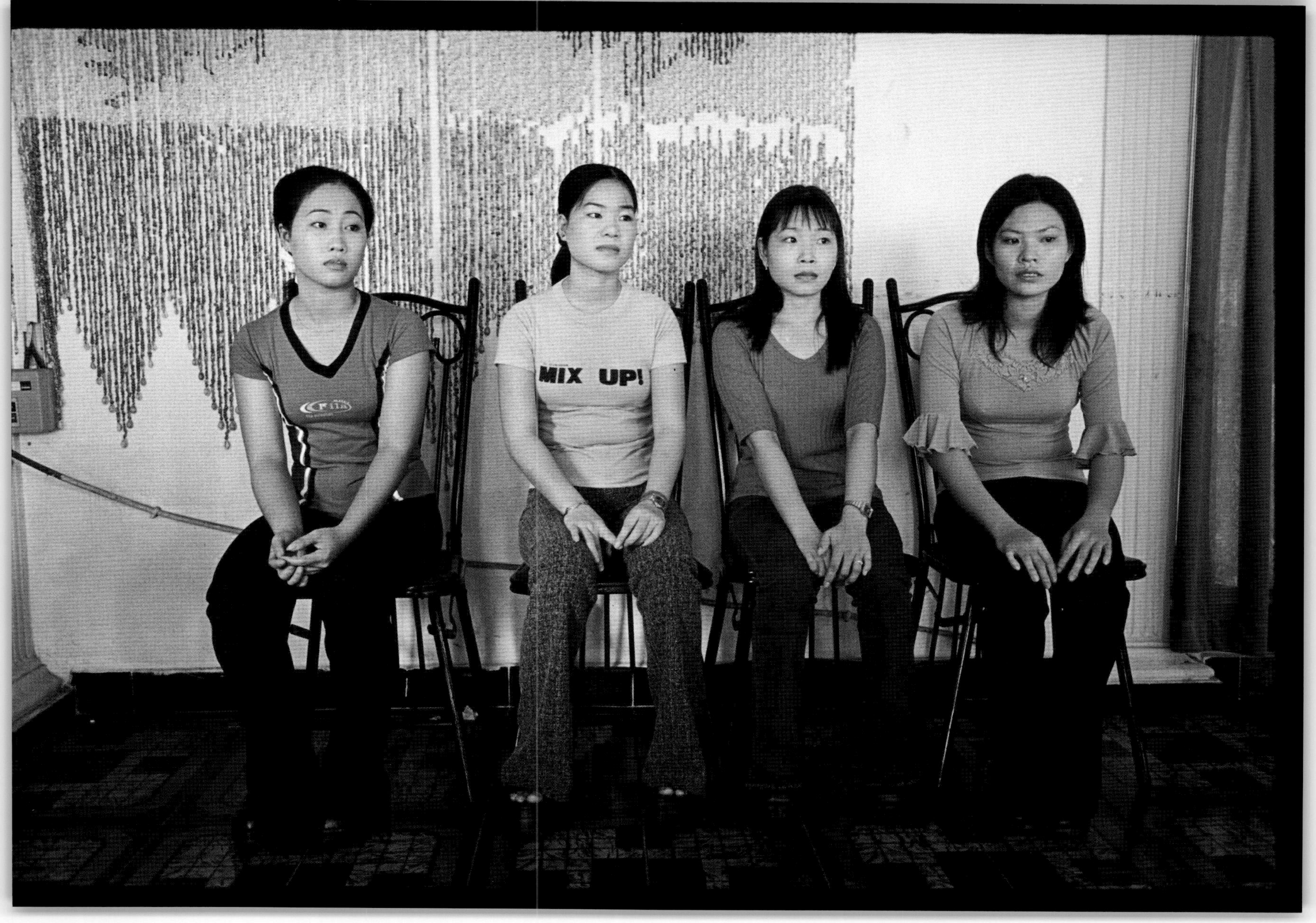
MIX UP!

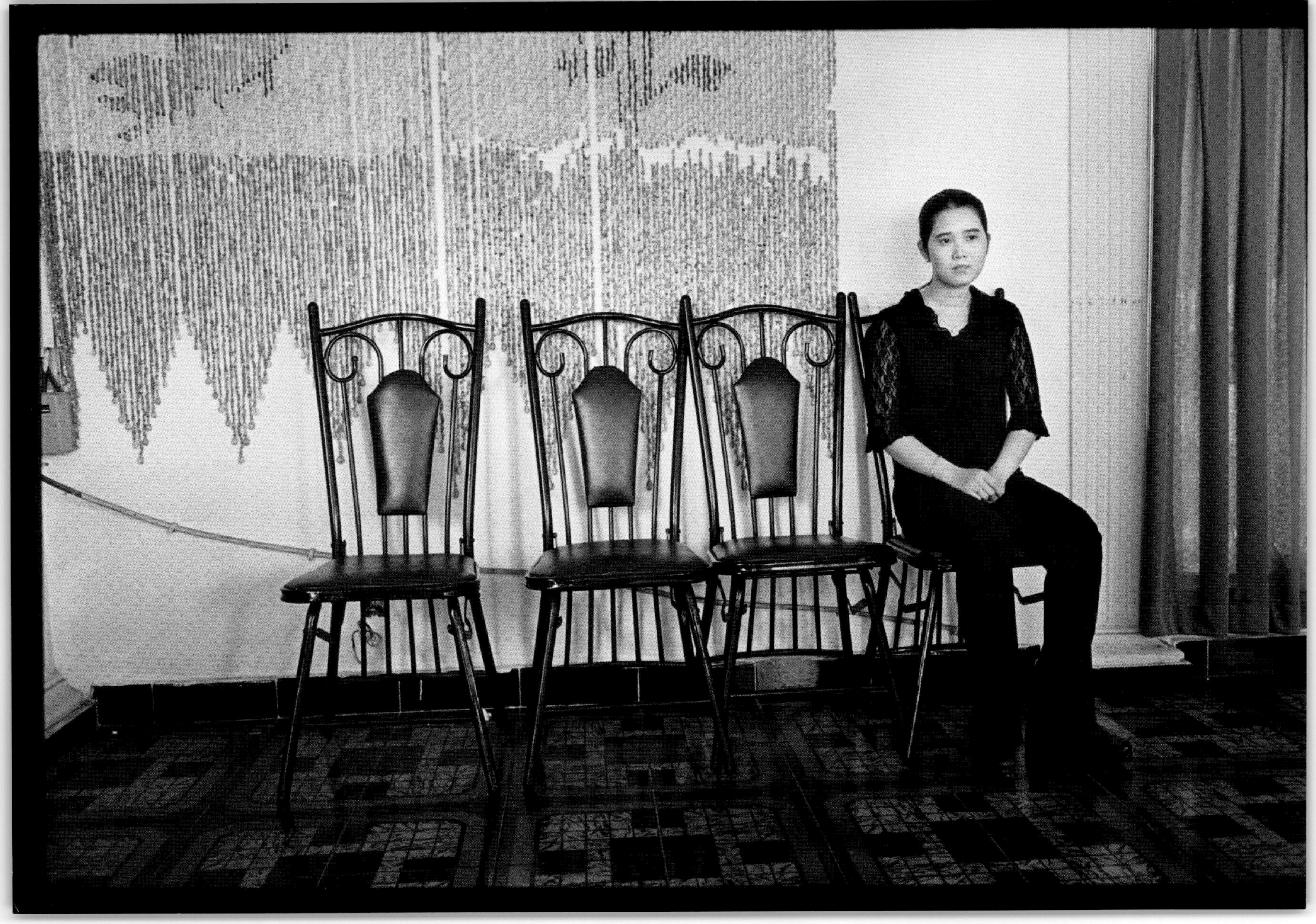

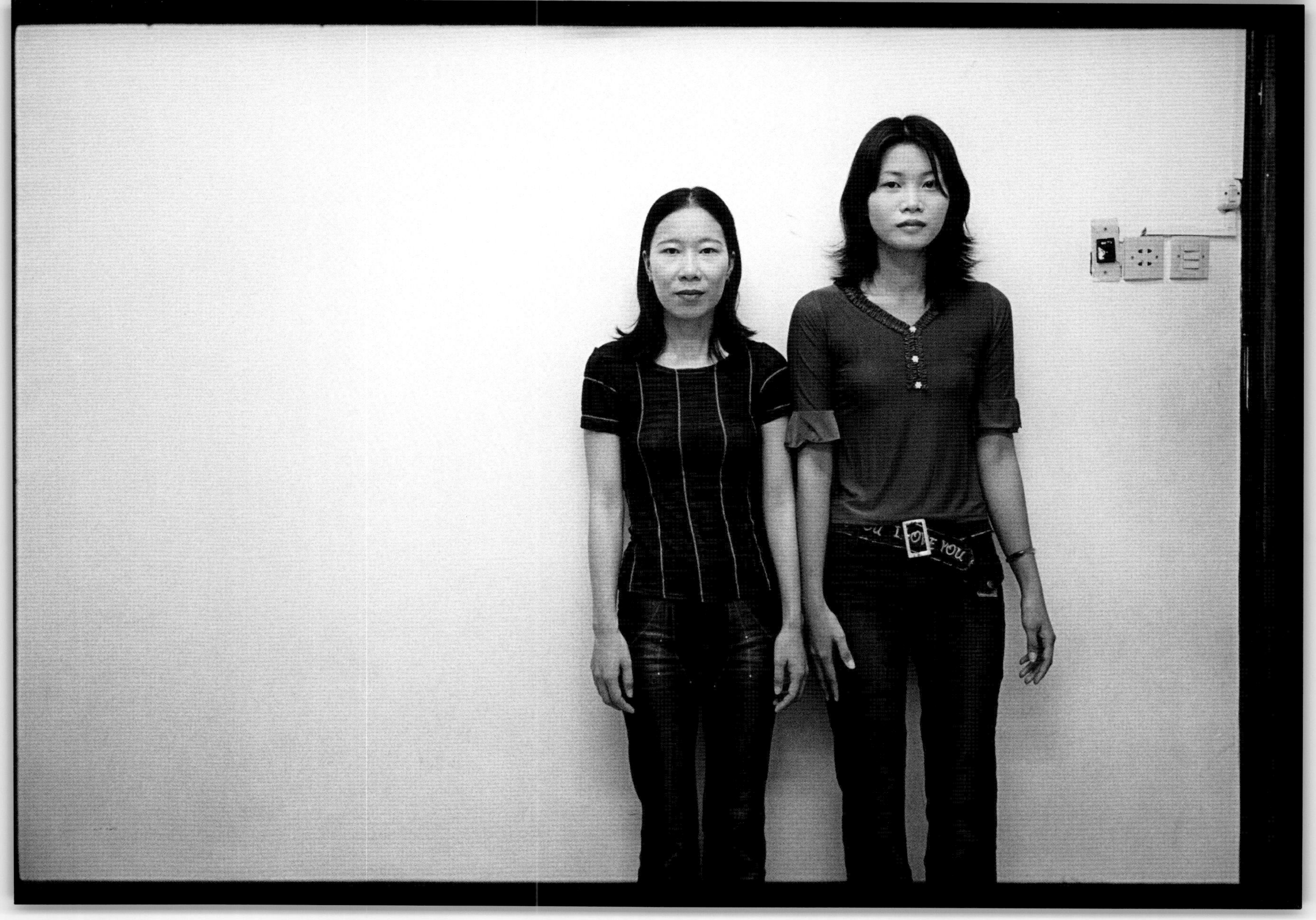

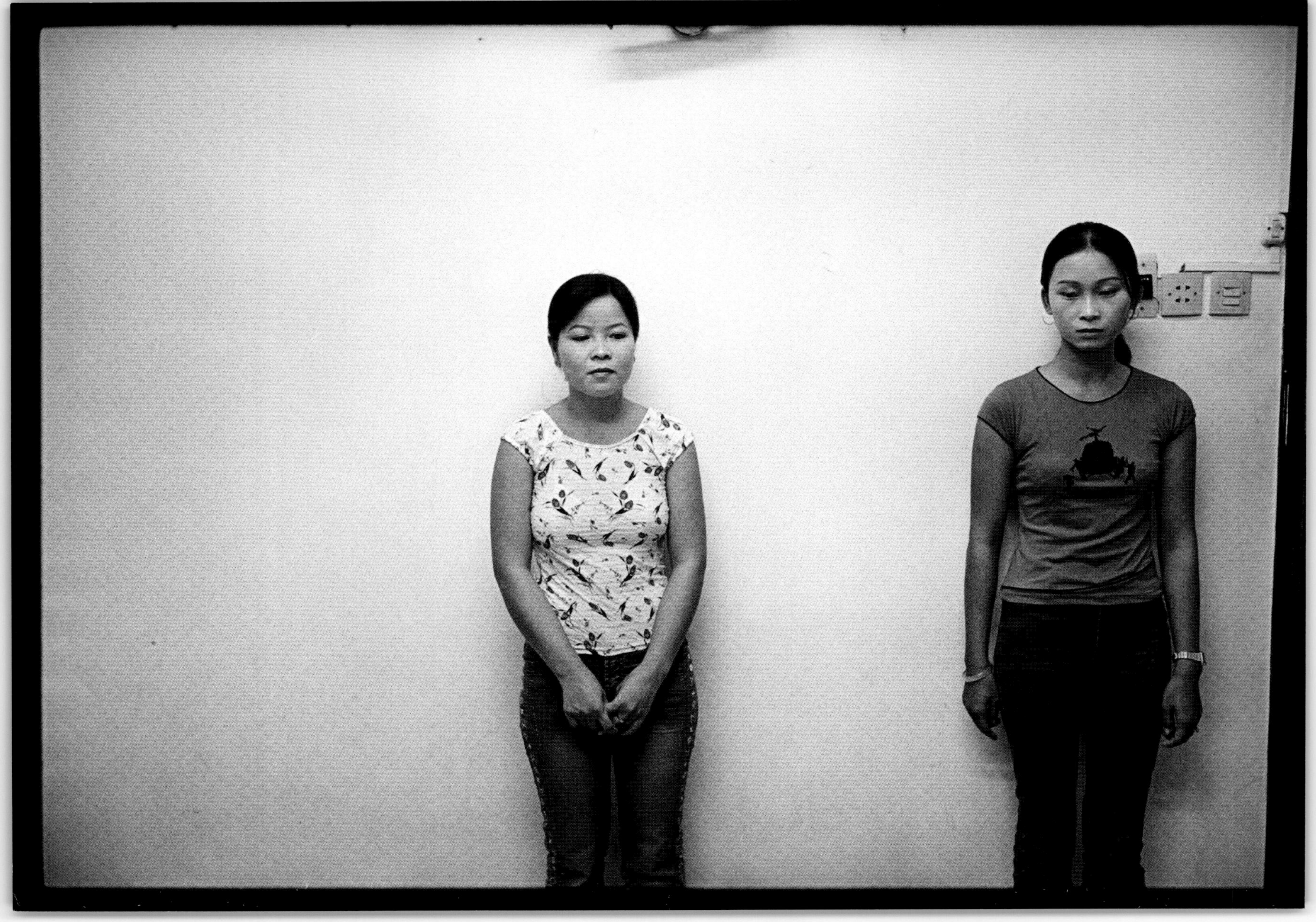

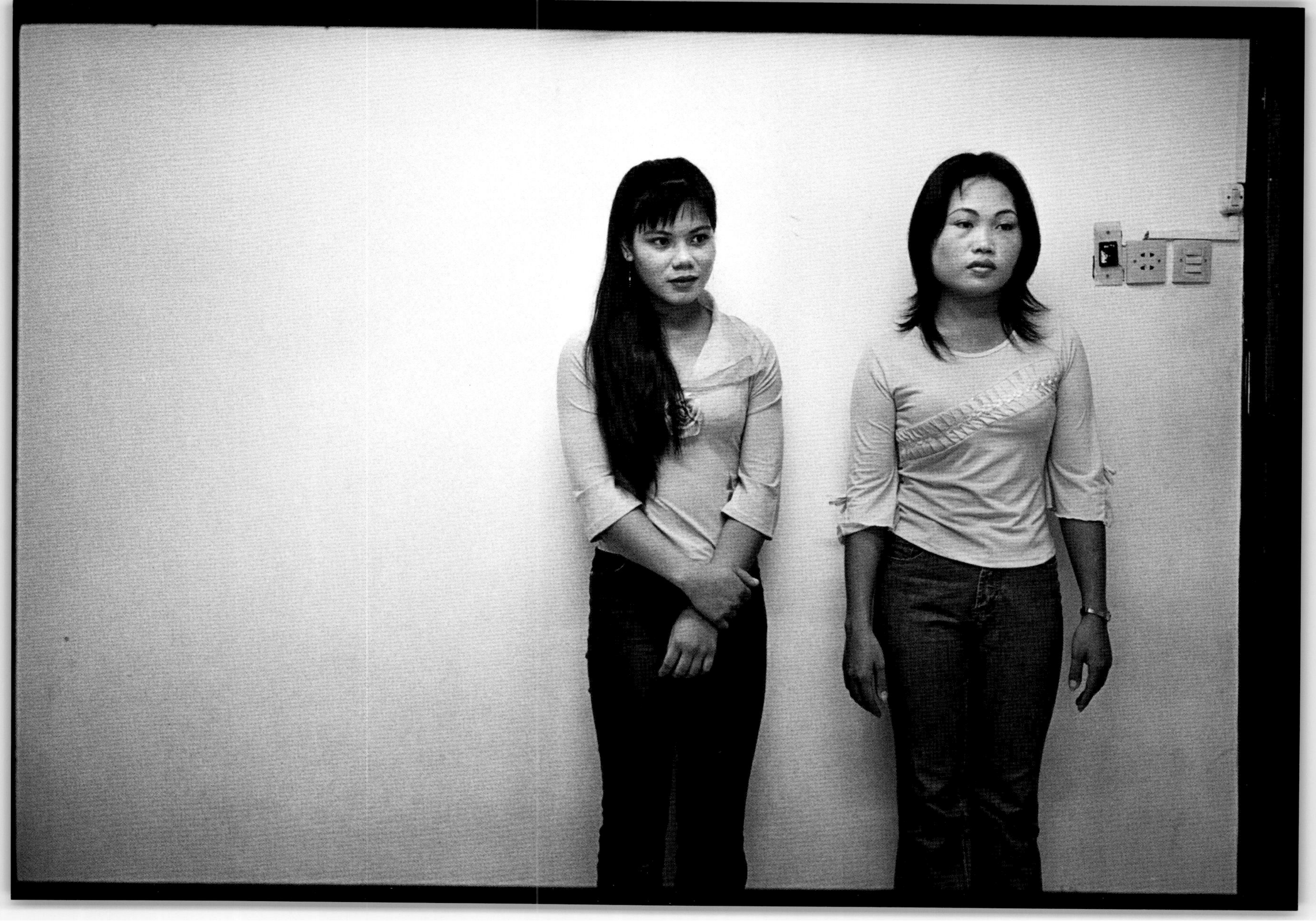

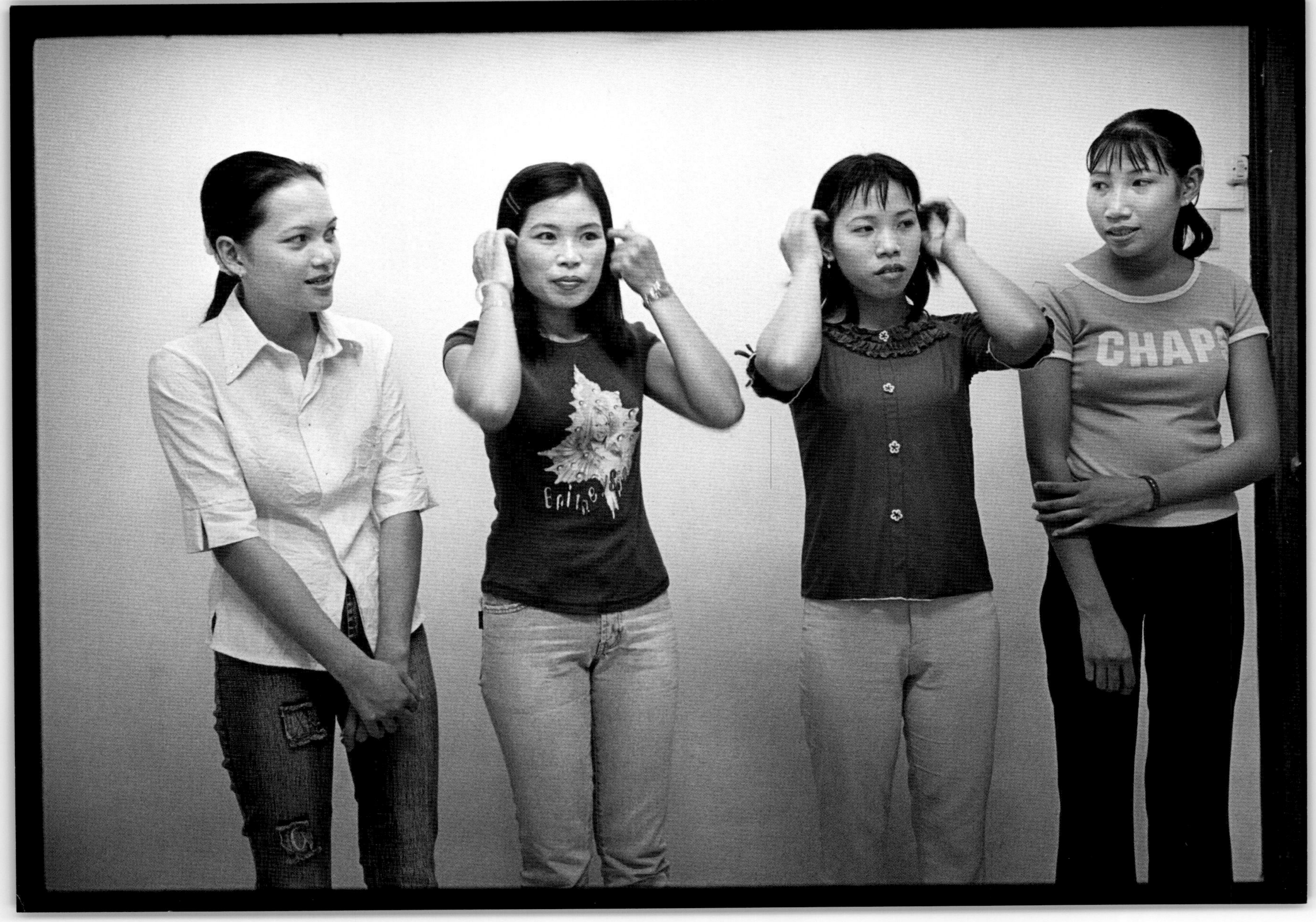
CHAP

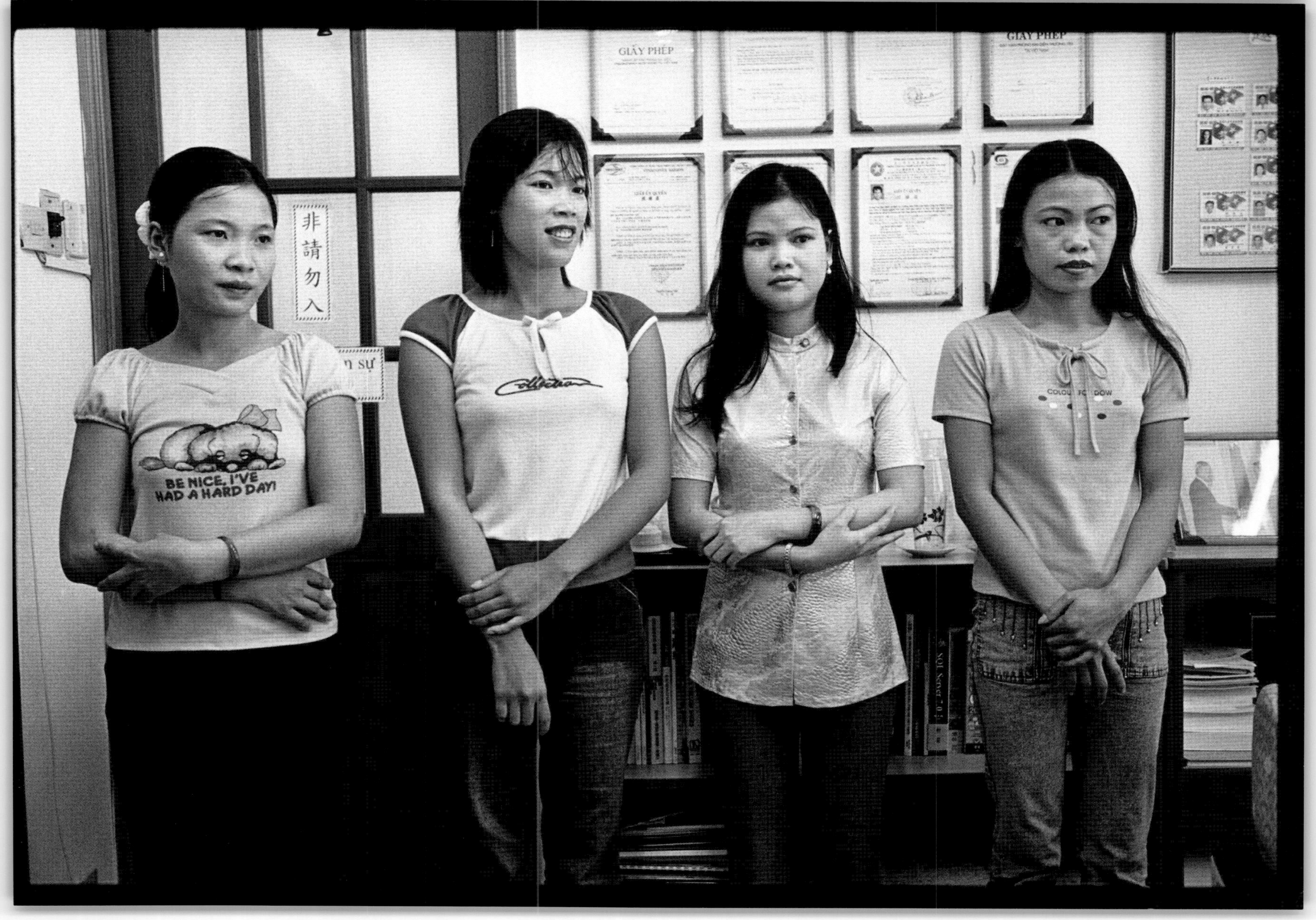

GIẤY PHÉP
GIẤY PHÉP
非請勿入
BE NICE, I'VE
HAD A HARD DAY!

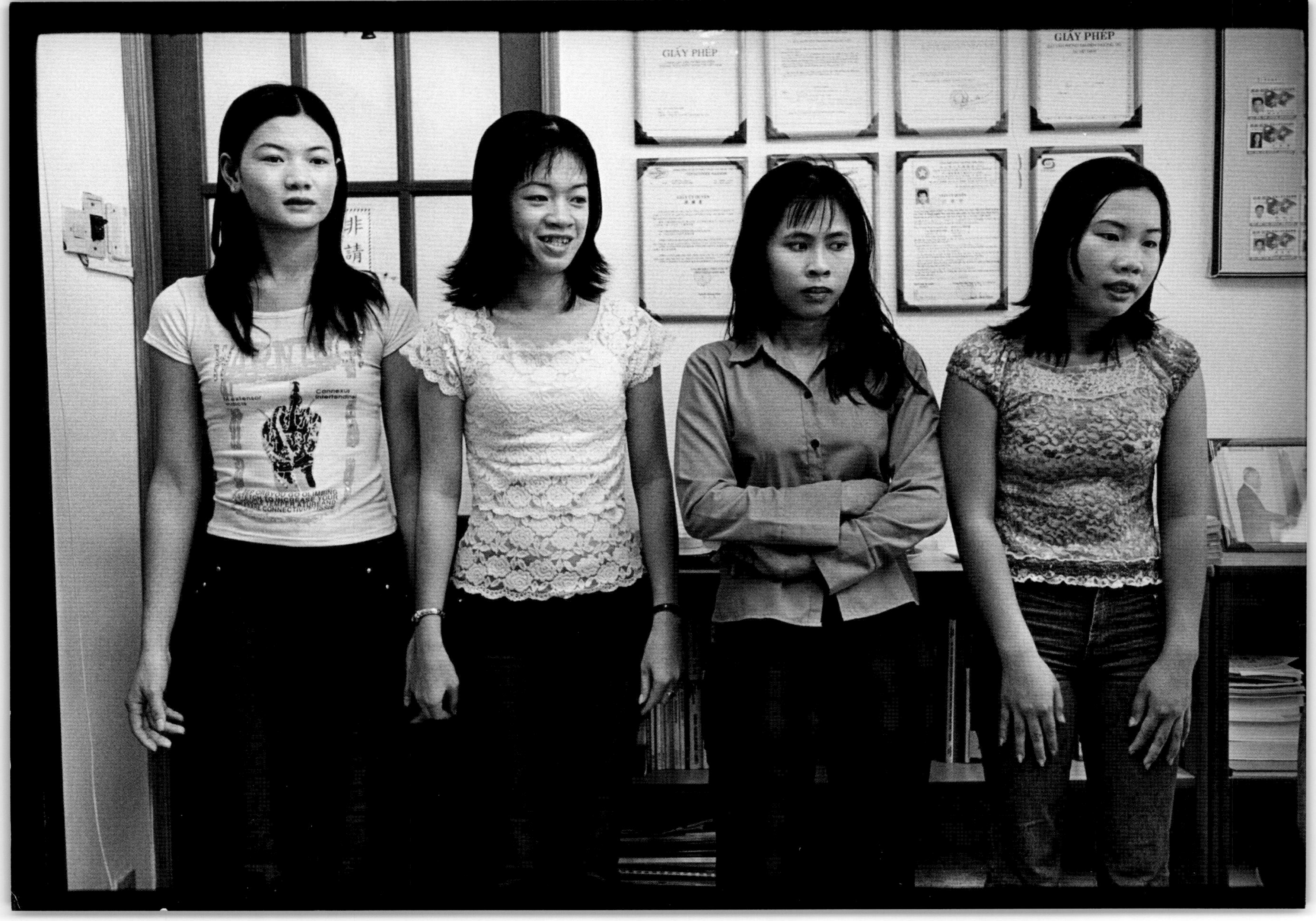
GIẤY PHÉP
GIẤY PHÉP

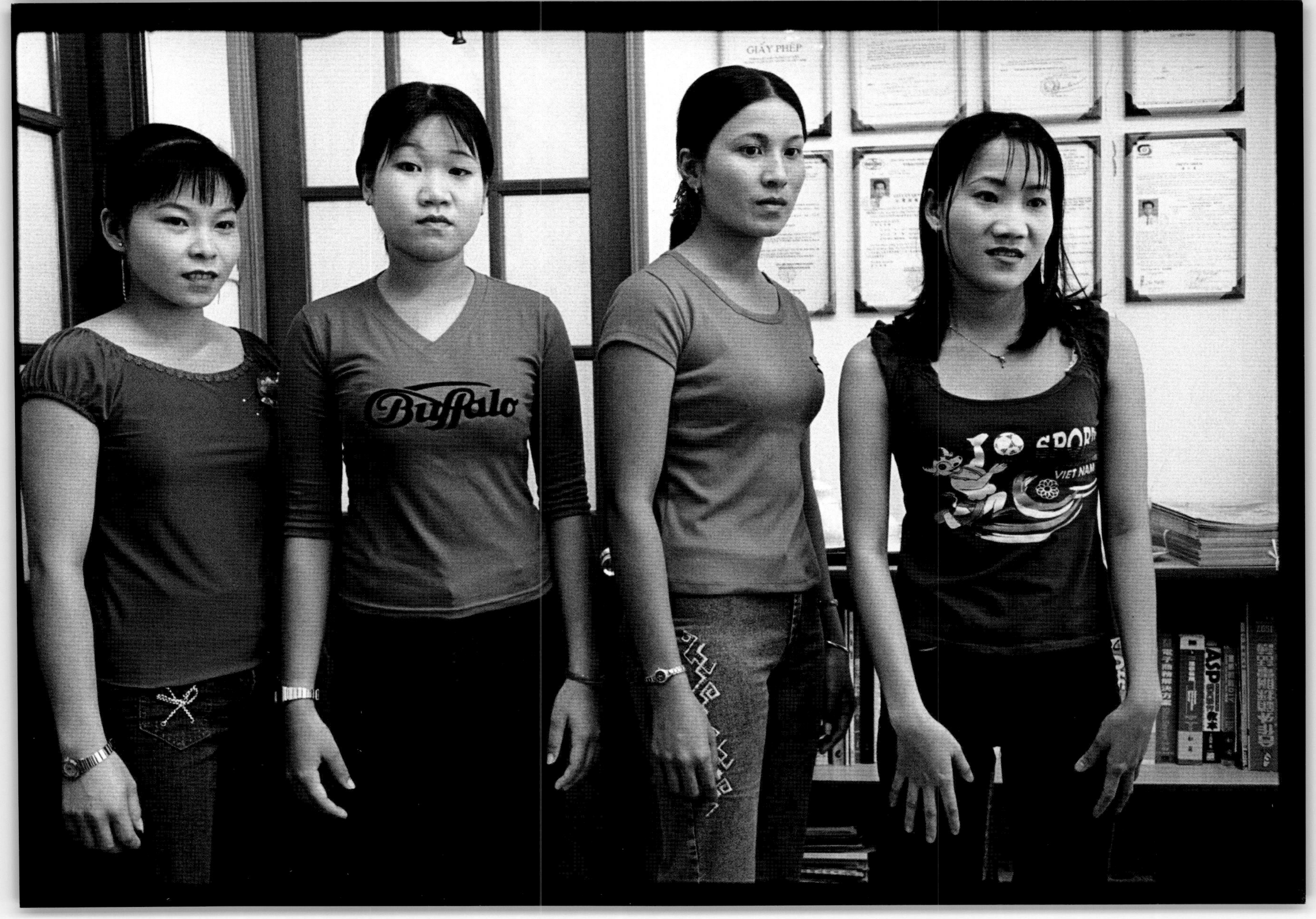
GIẤY PHÉP
Buffalo
VIET NAM

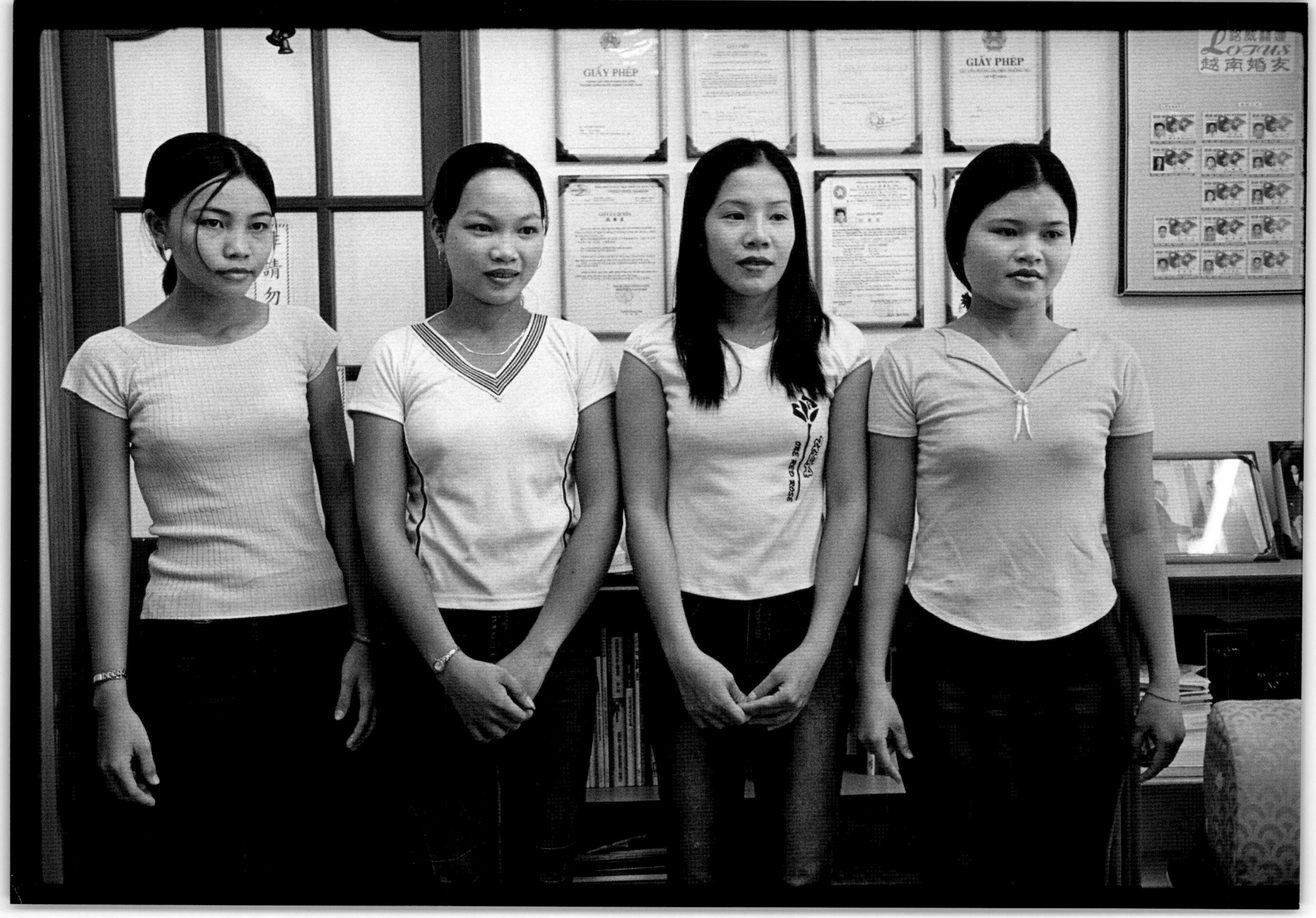
GIẤY PHÉP
GIẤY PHÉP
LOTUS
越南婚友
ONE RED ROSE

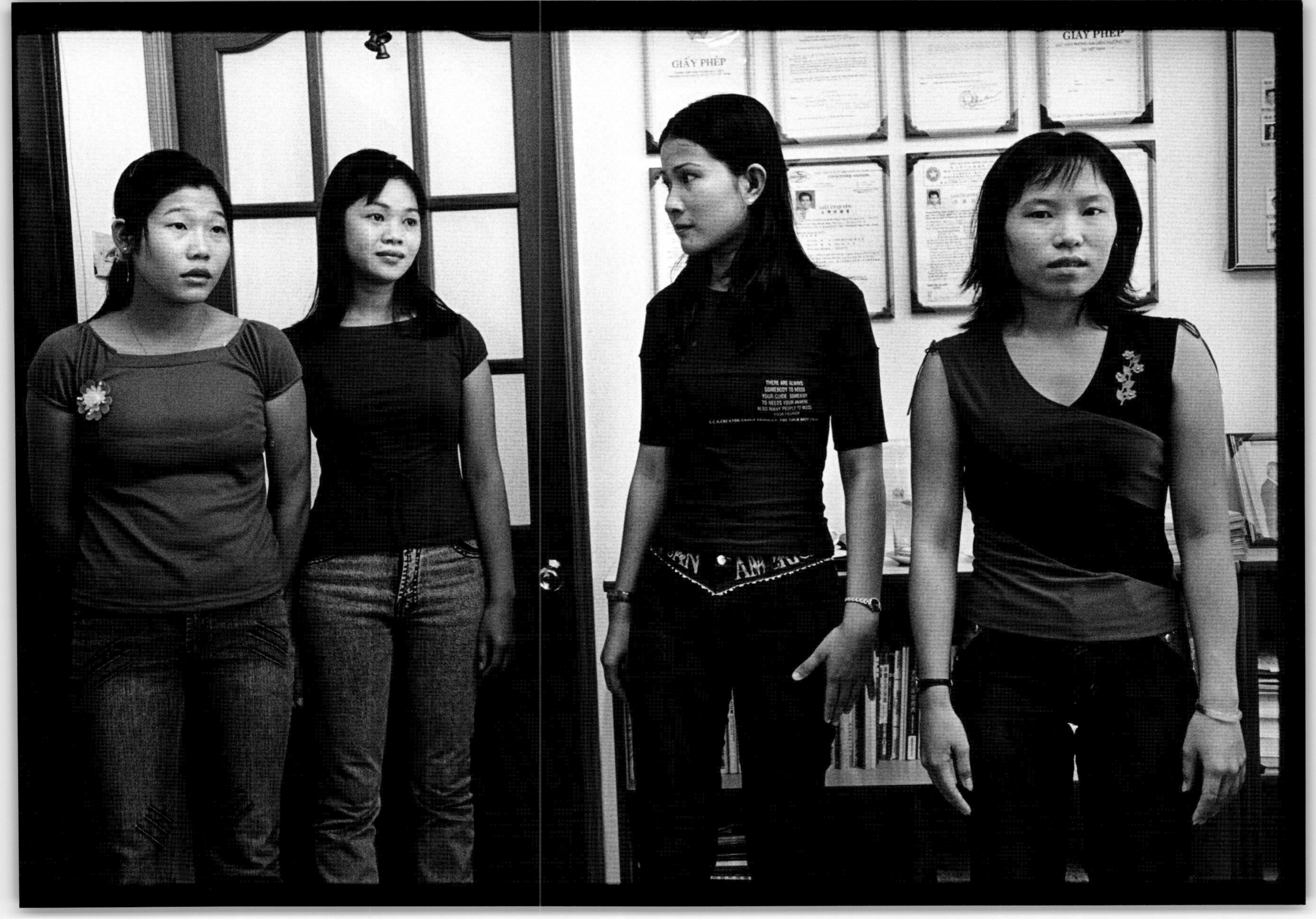
GIẤY PHÉP
GIẤY PHÉP

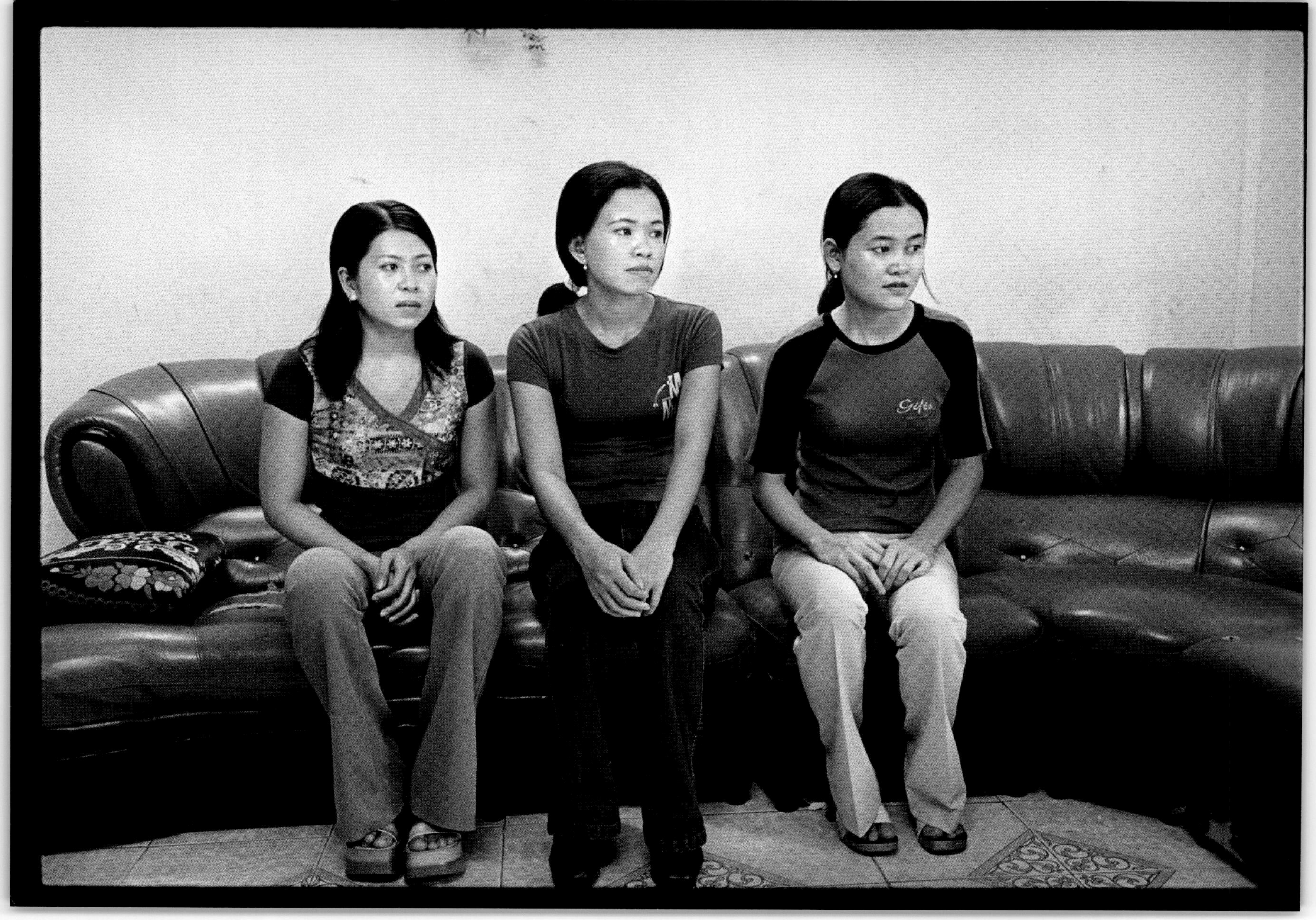

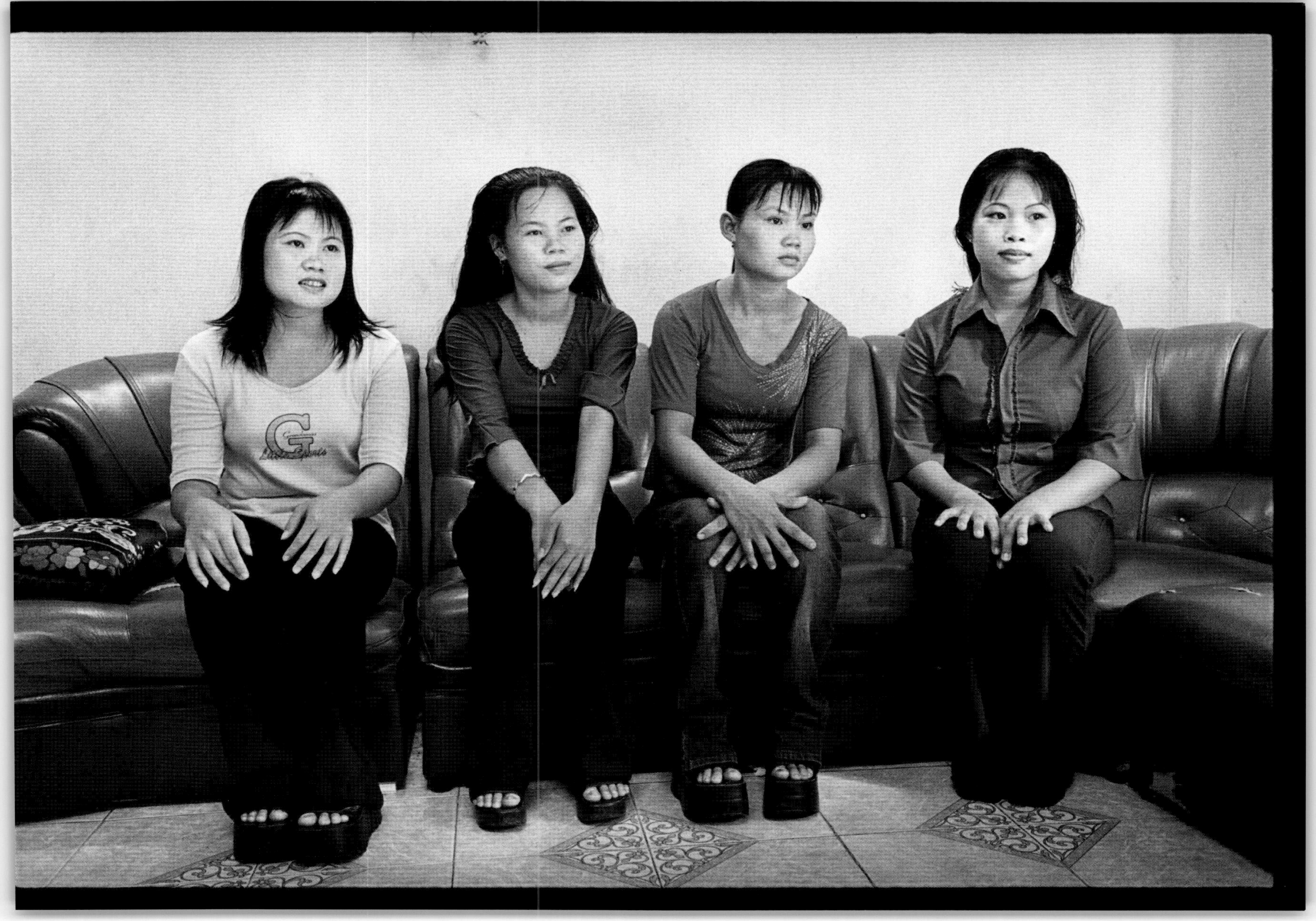

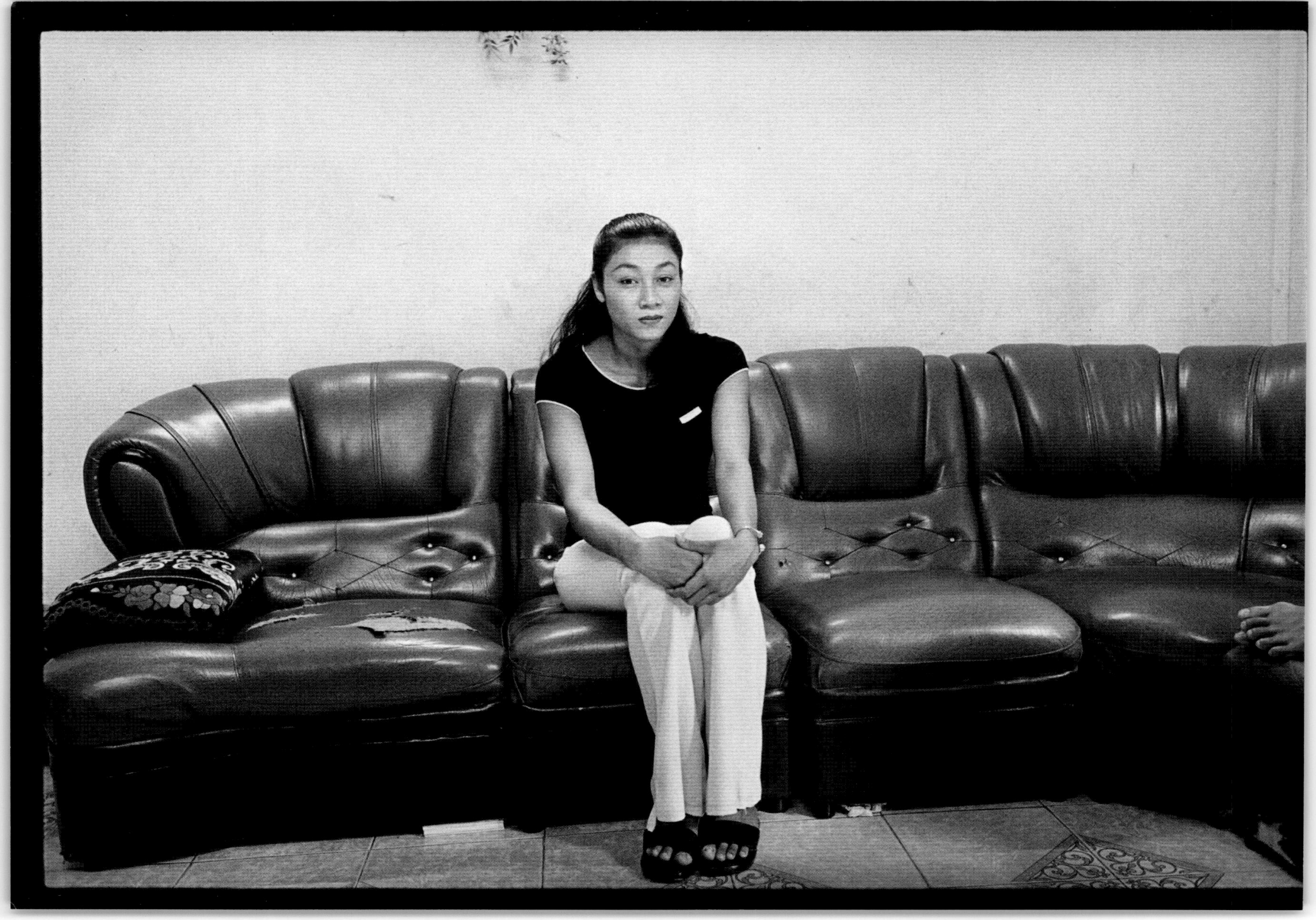

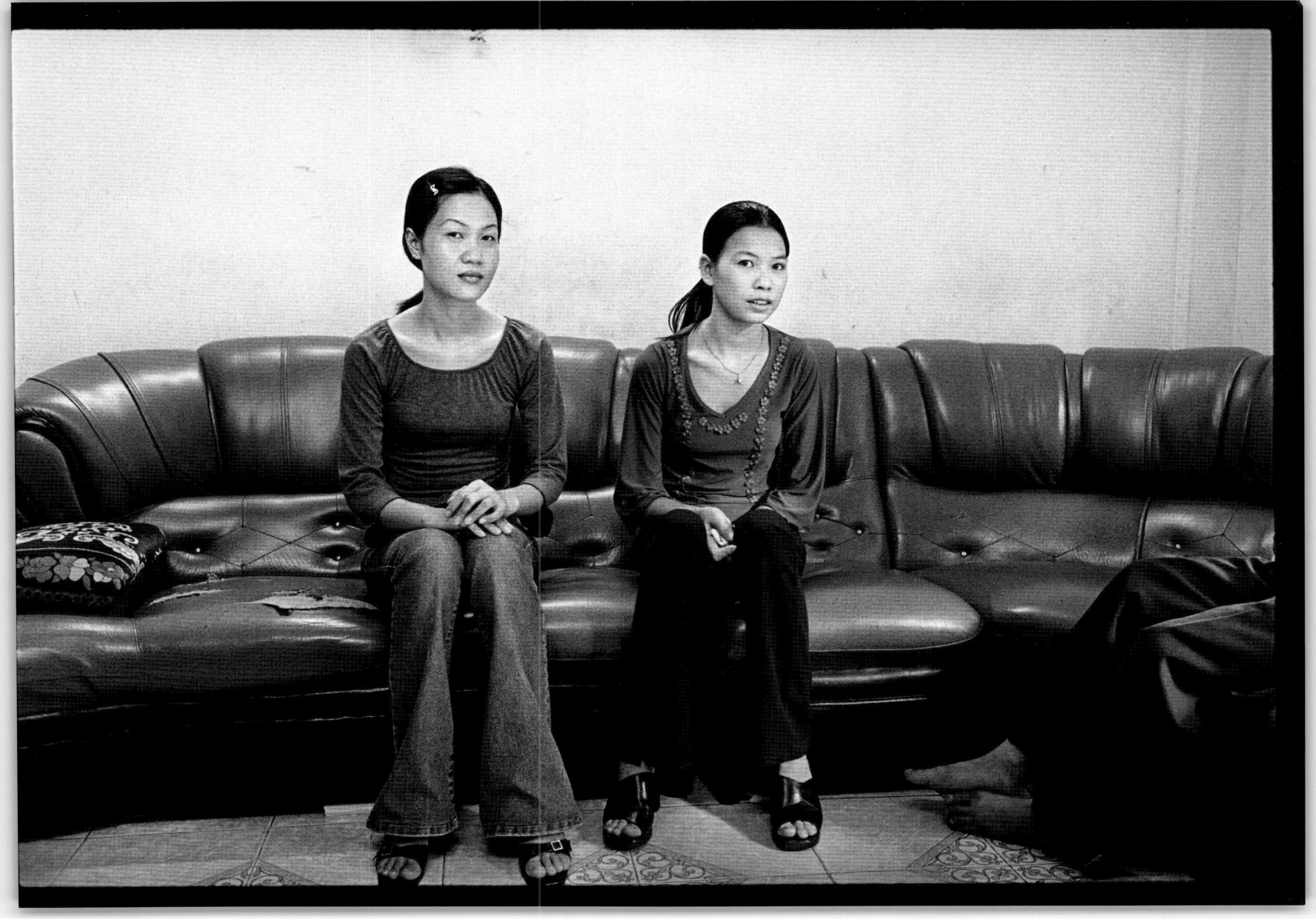

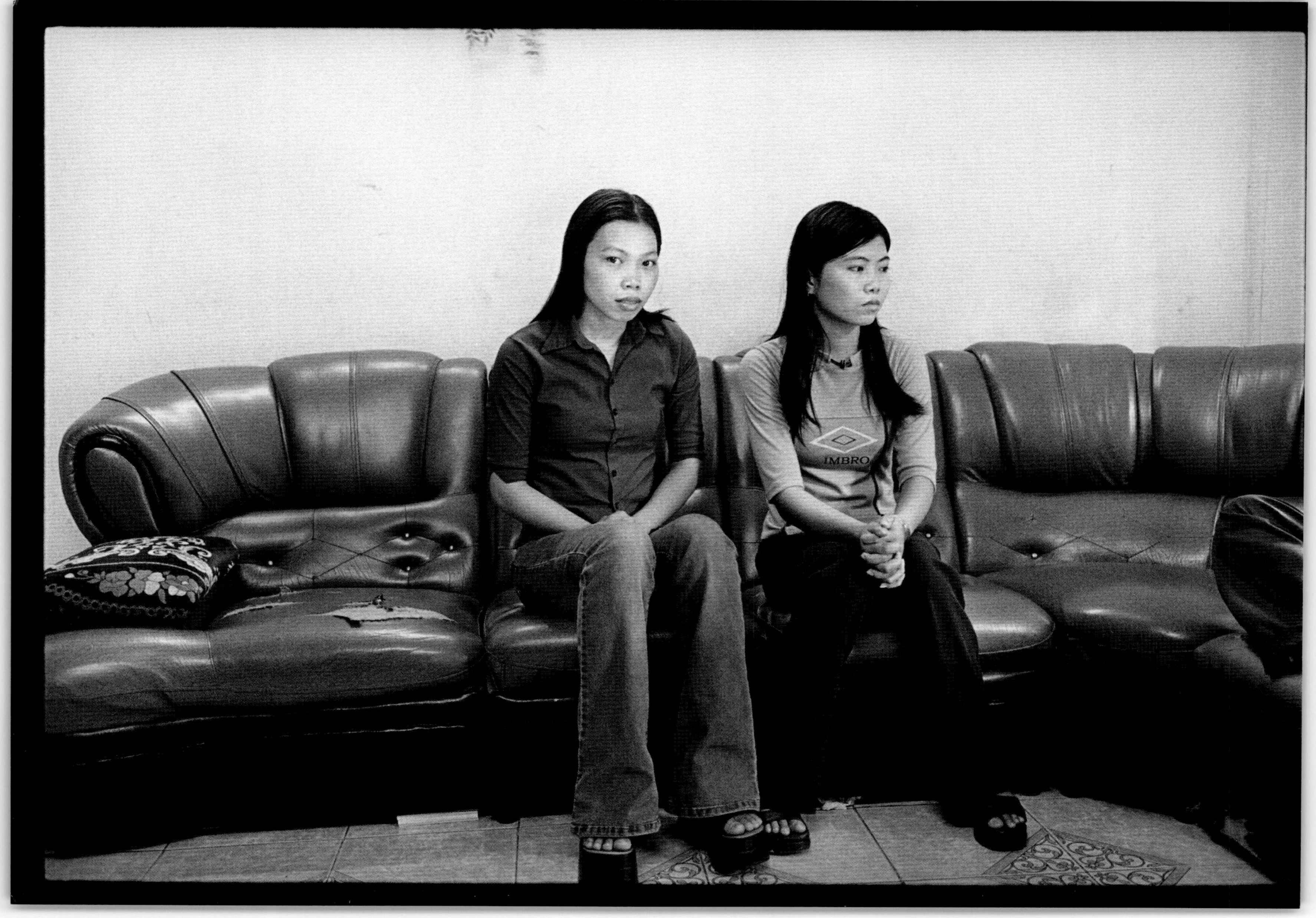
IMBRO

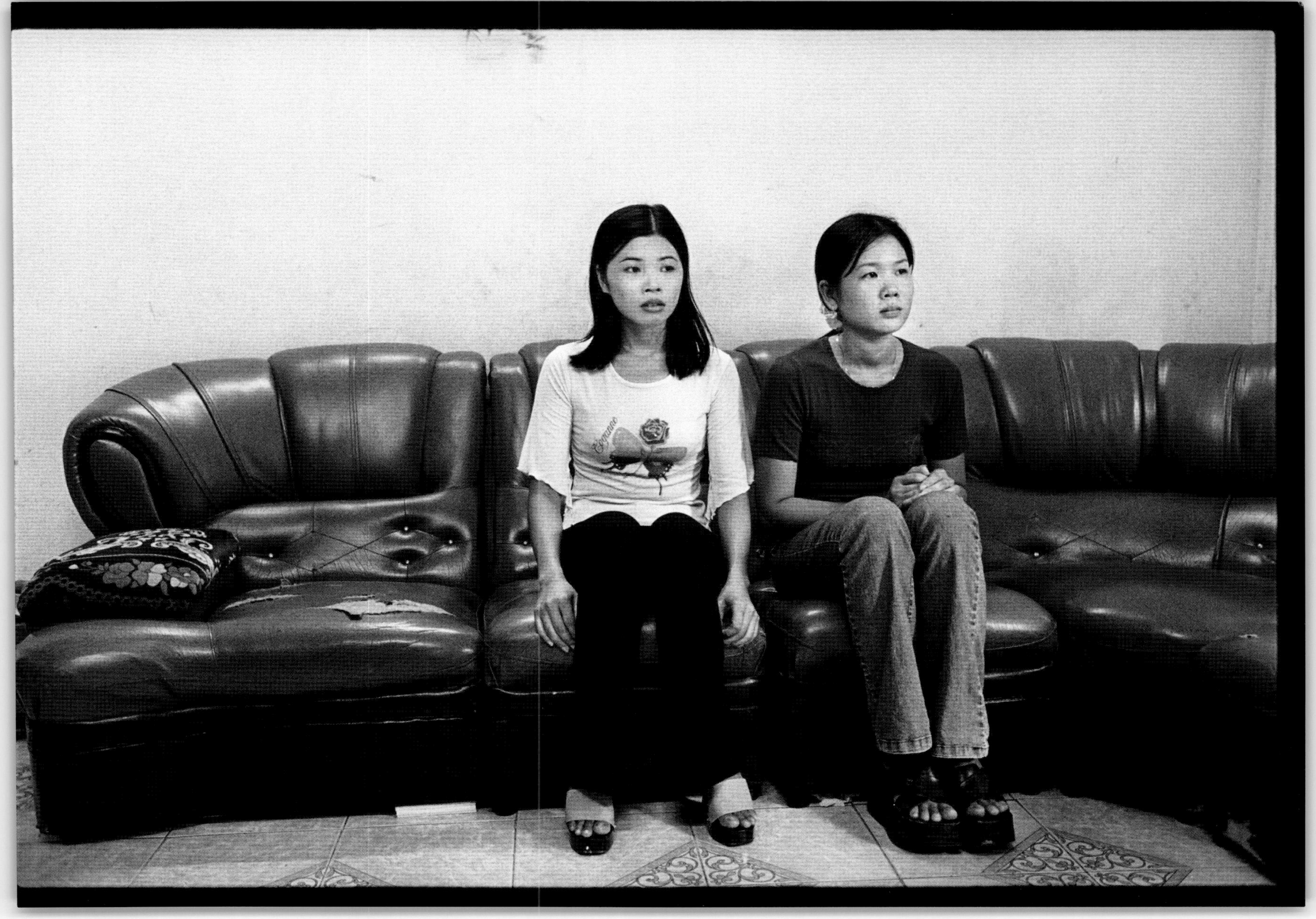

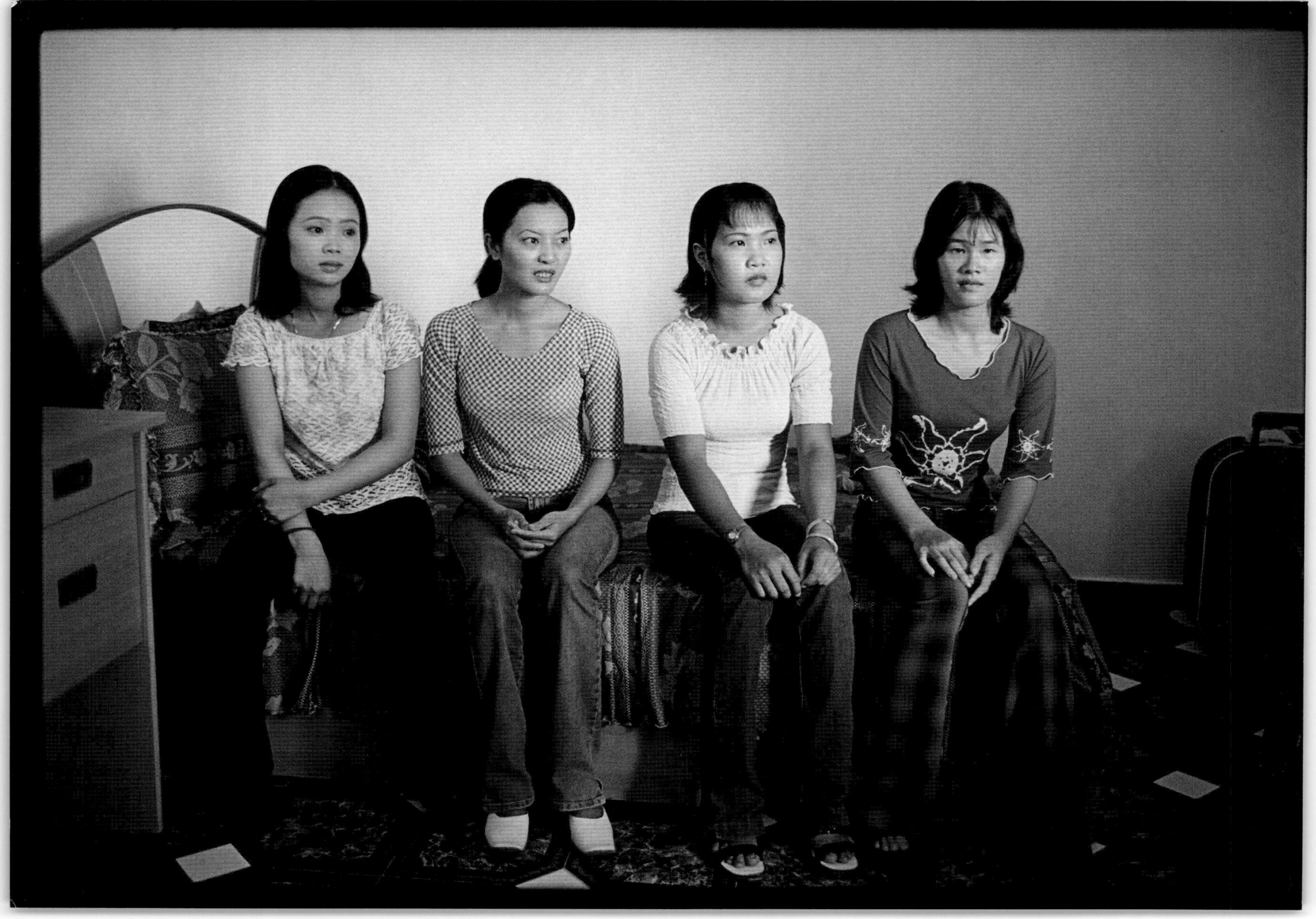

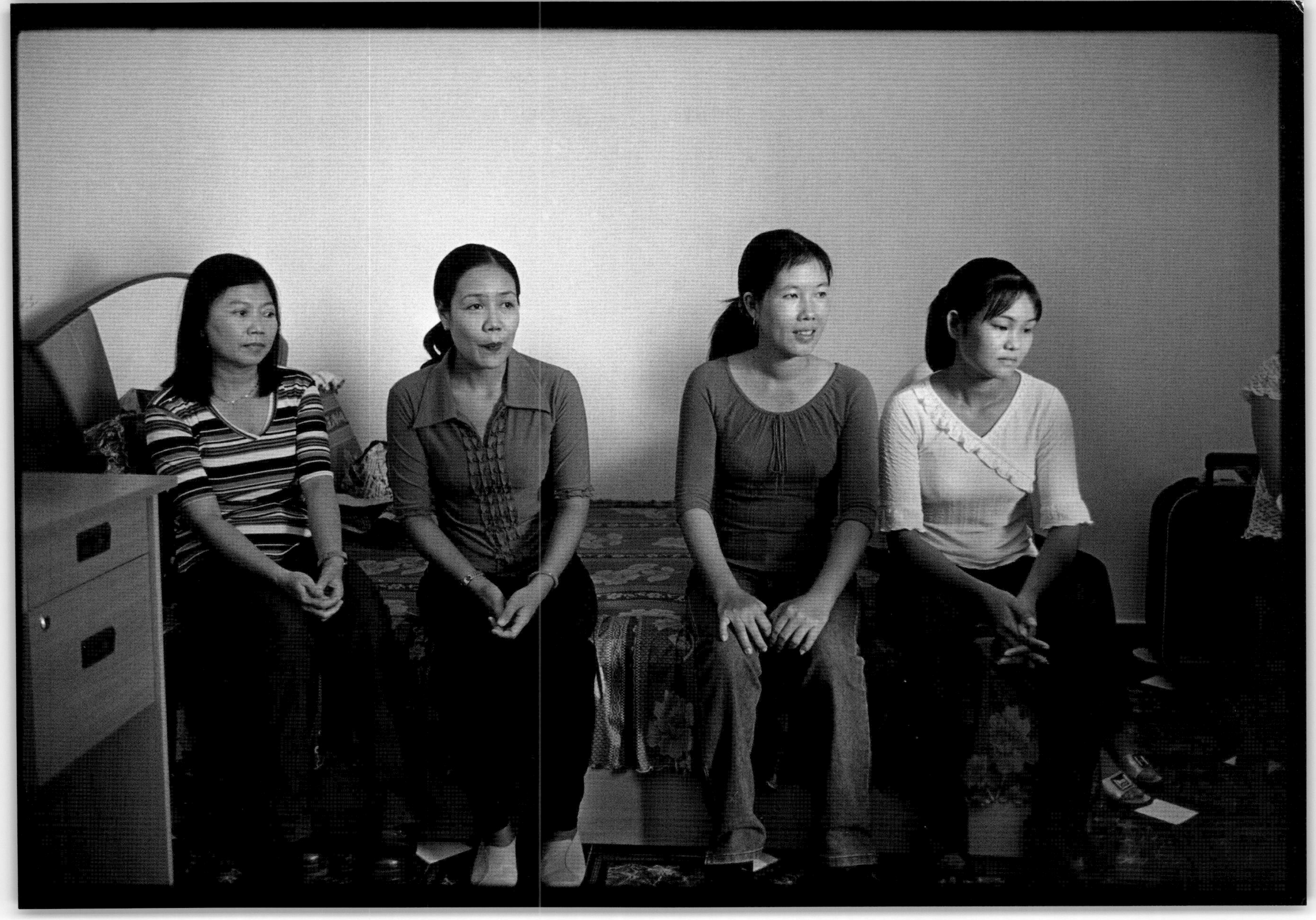

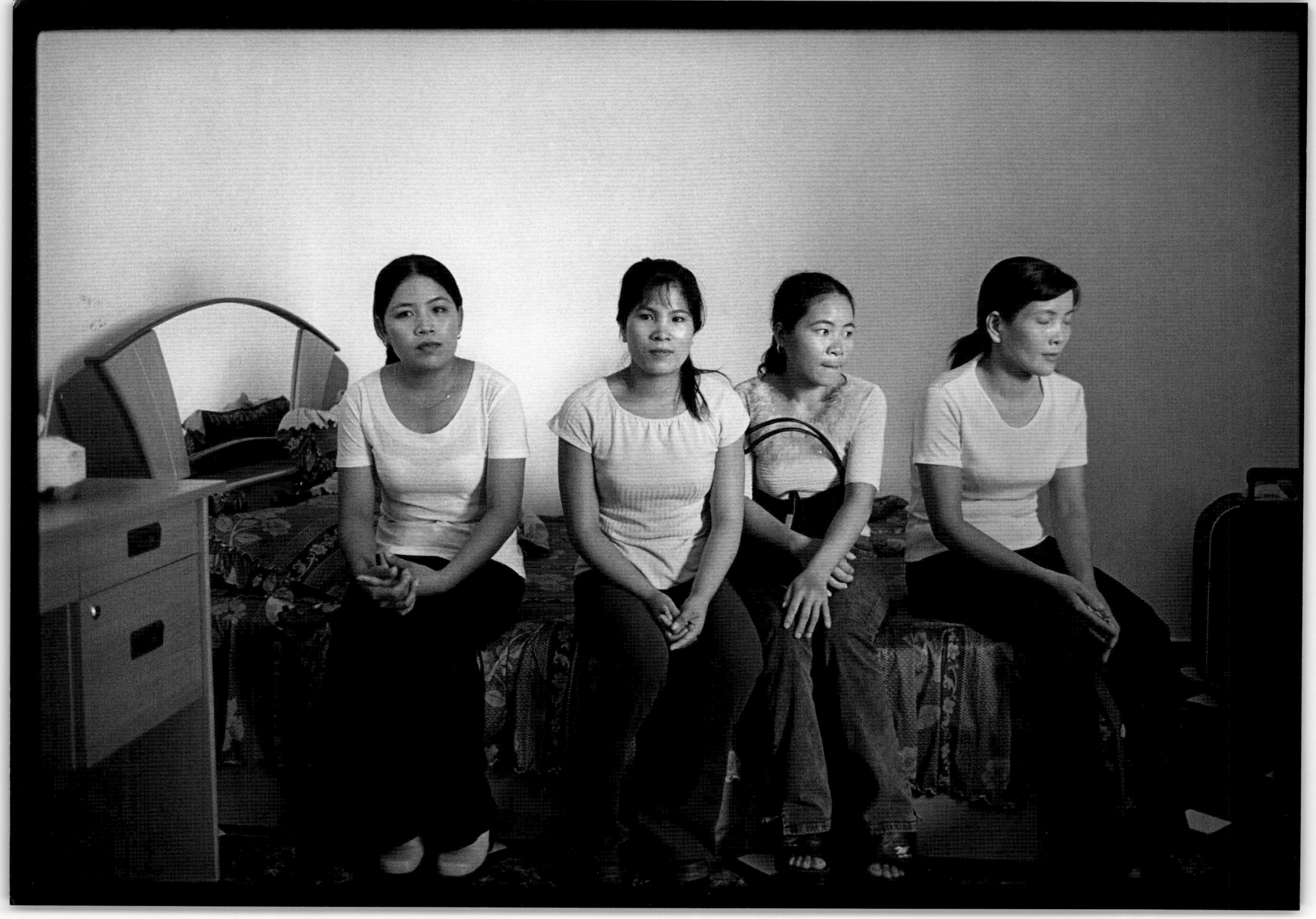

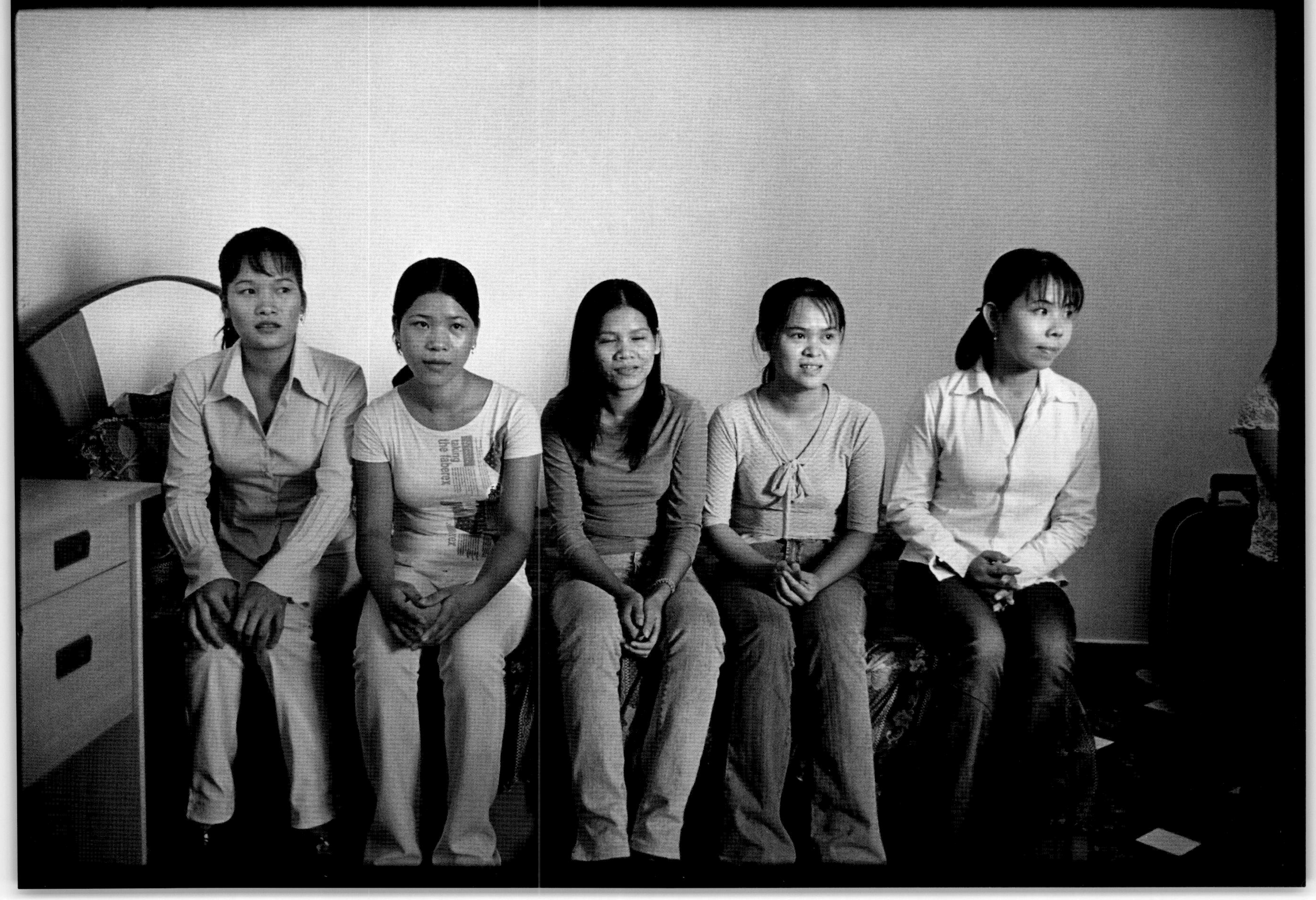

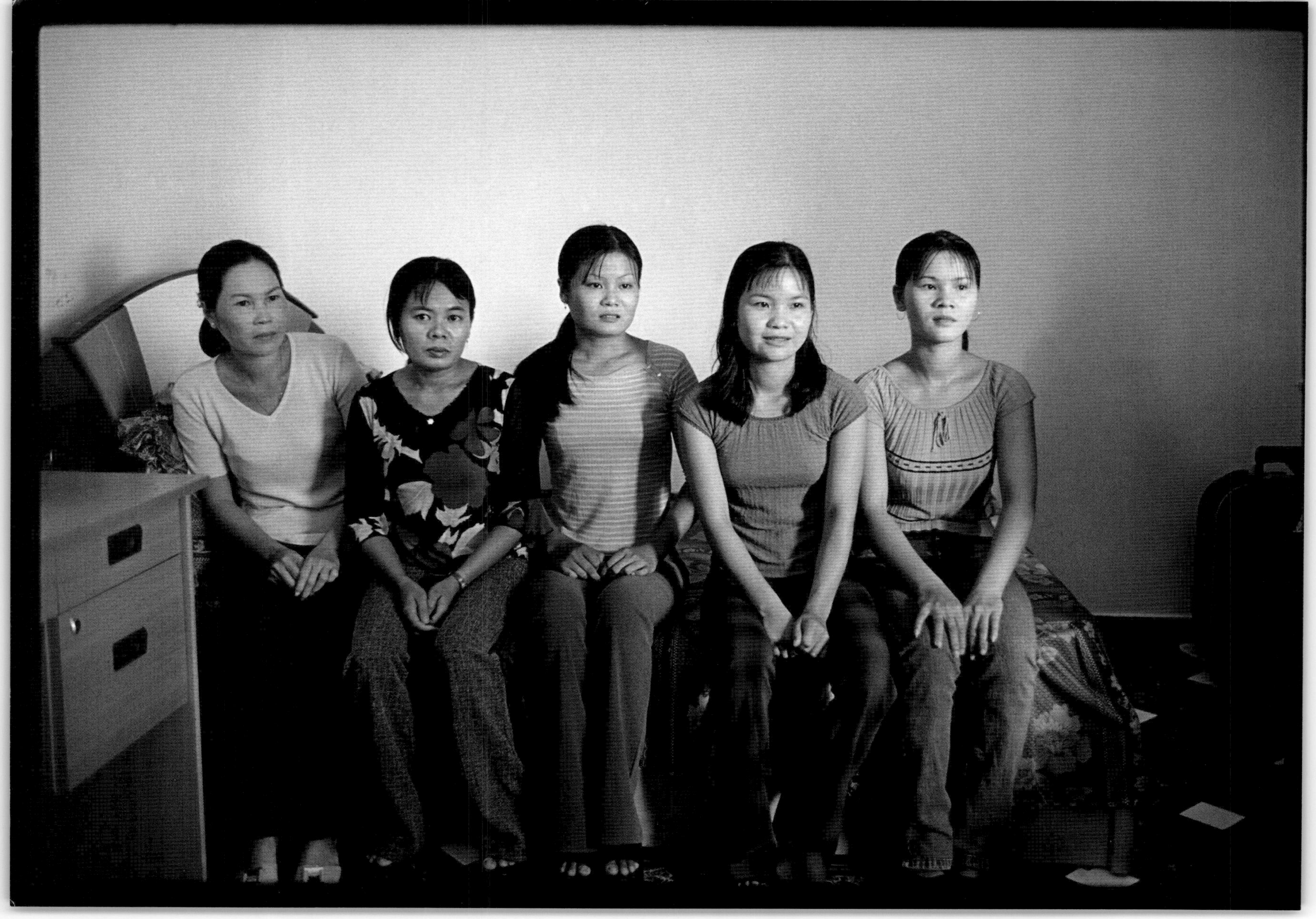

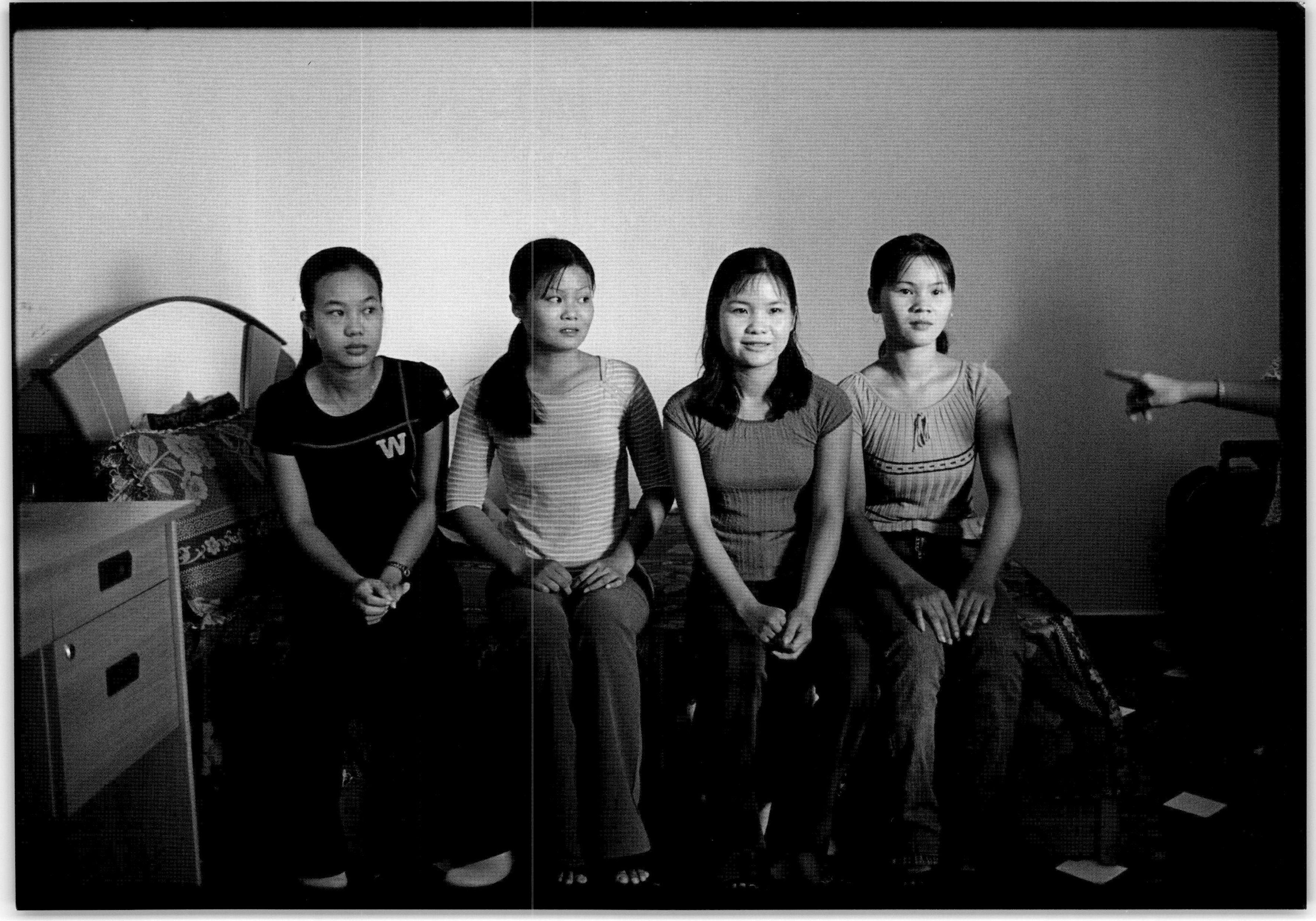

After the men have made their selection, couples must present proof of identity so that the women can later apply for visas to Taiwan.

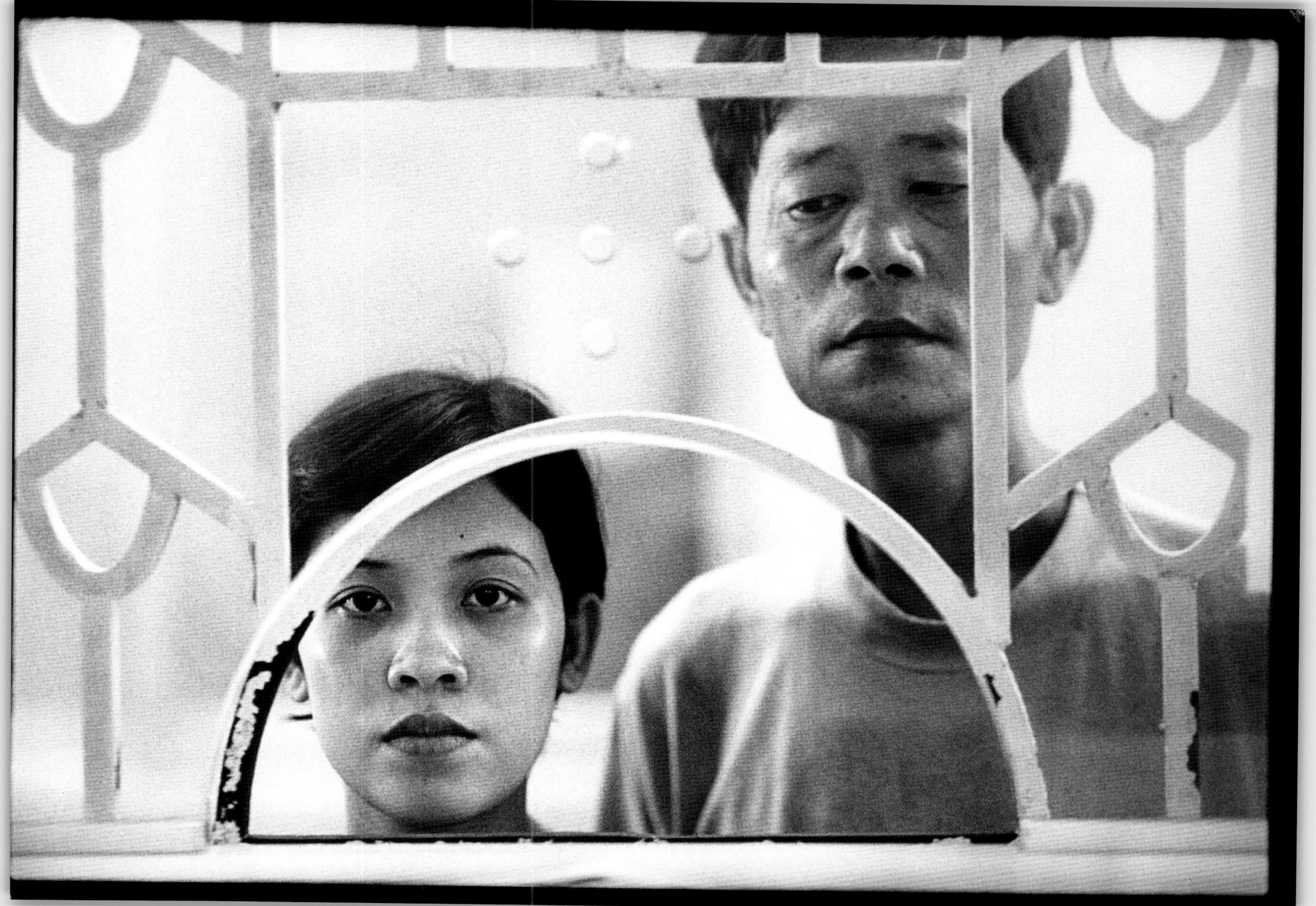

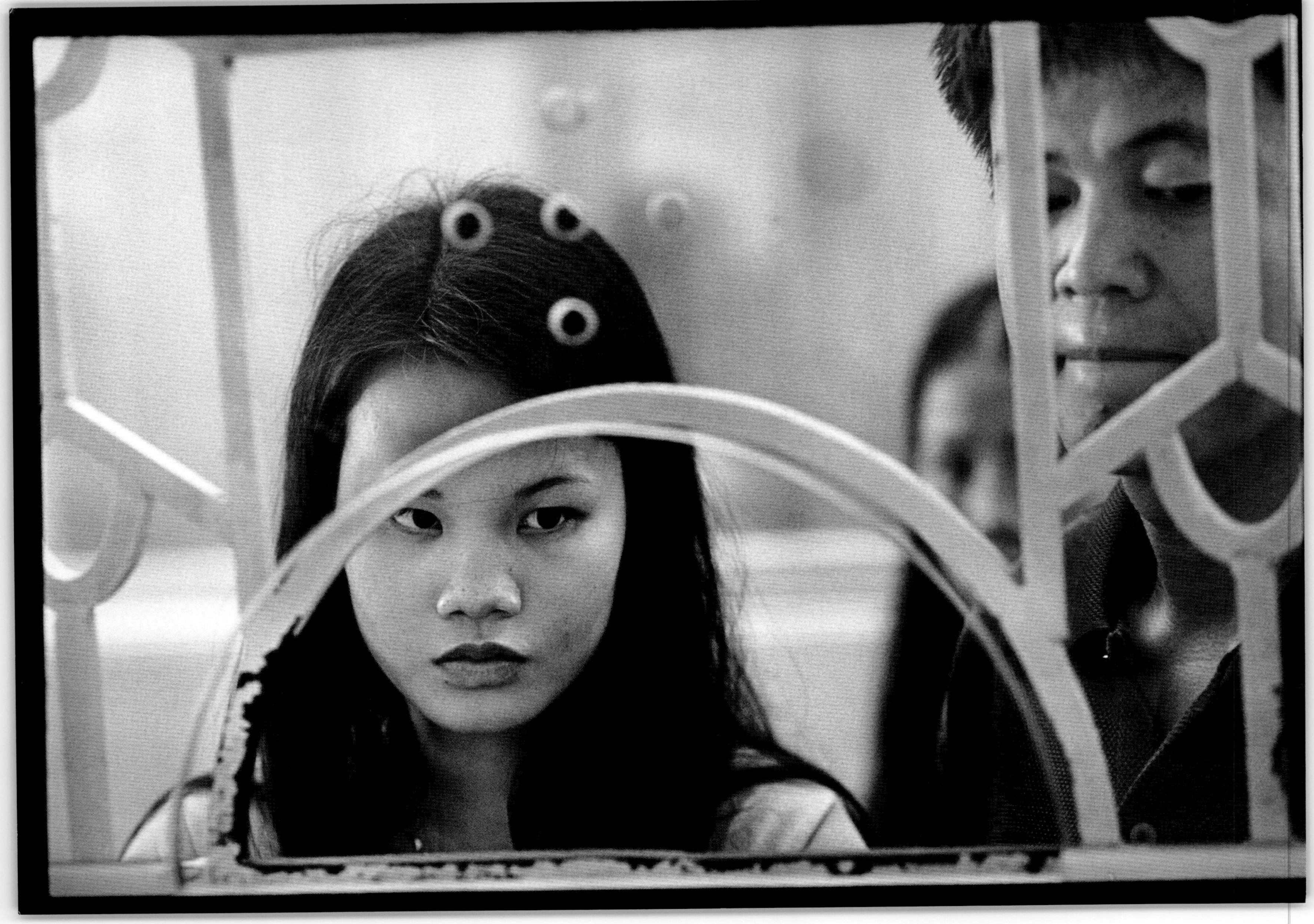

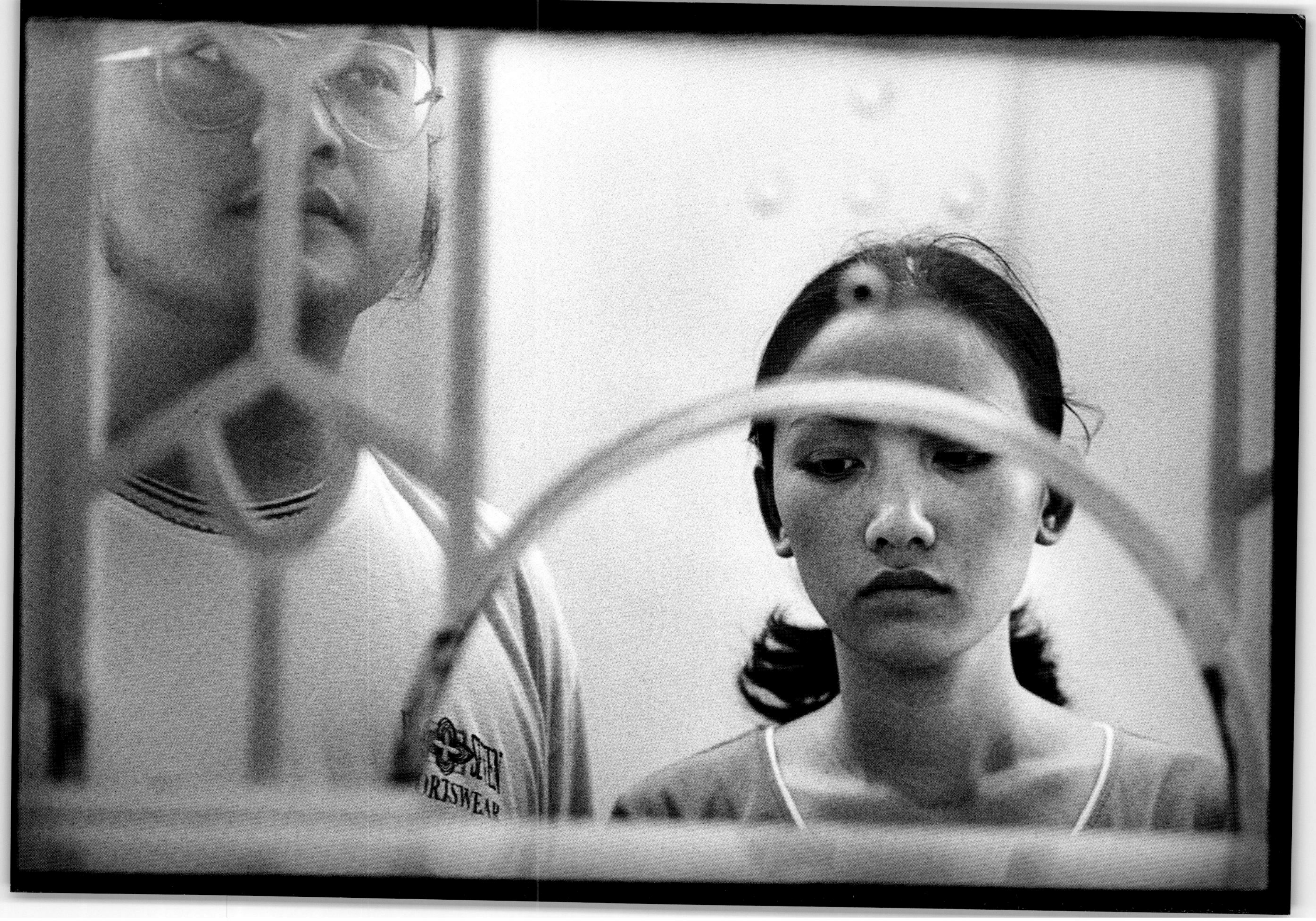

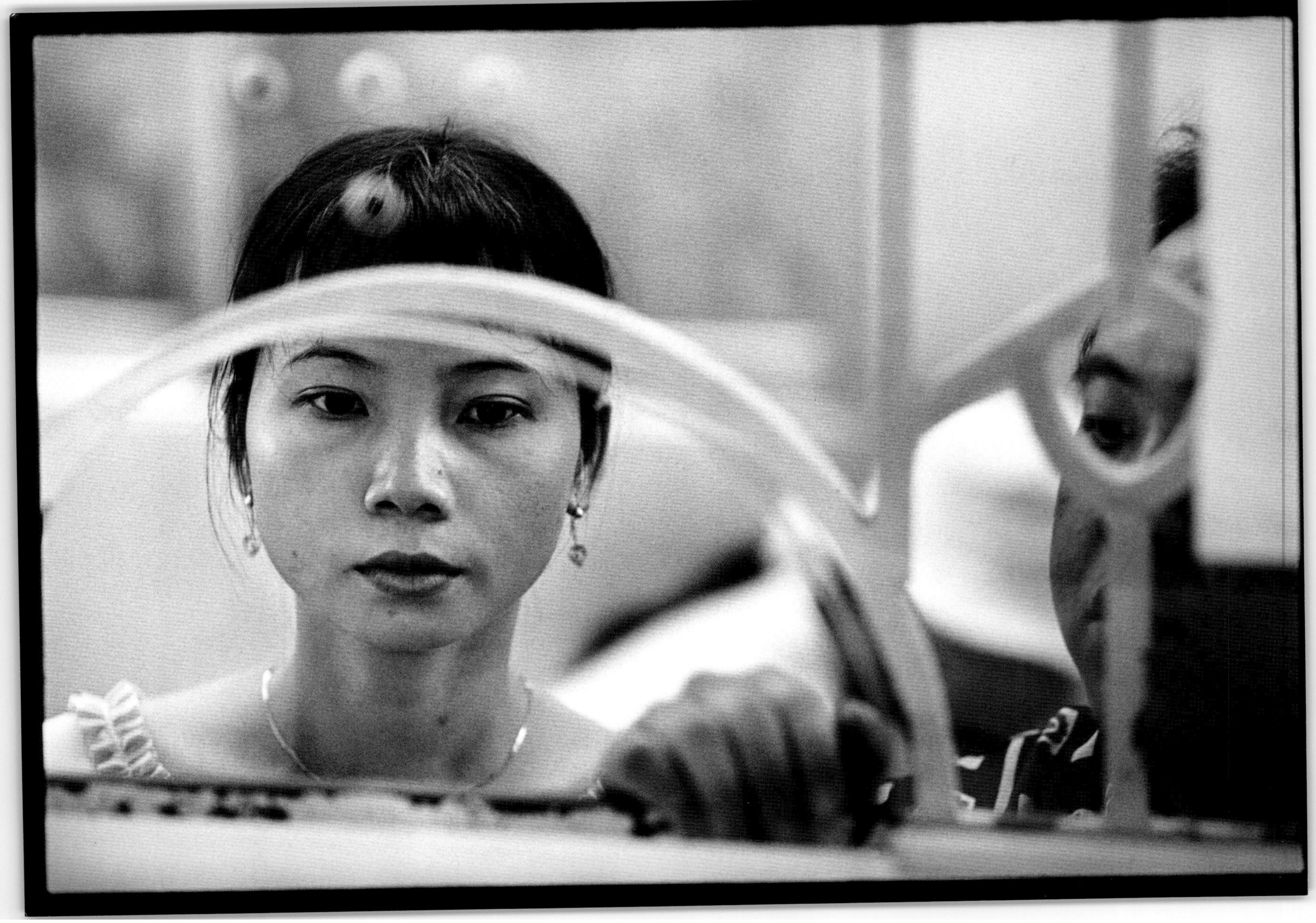

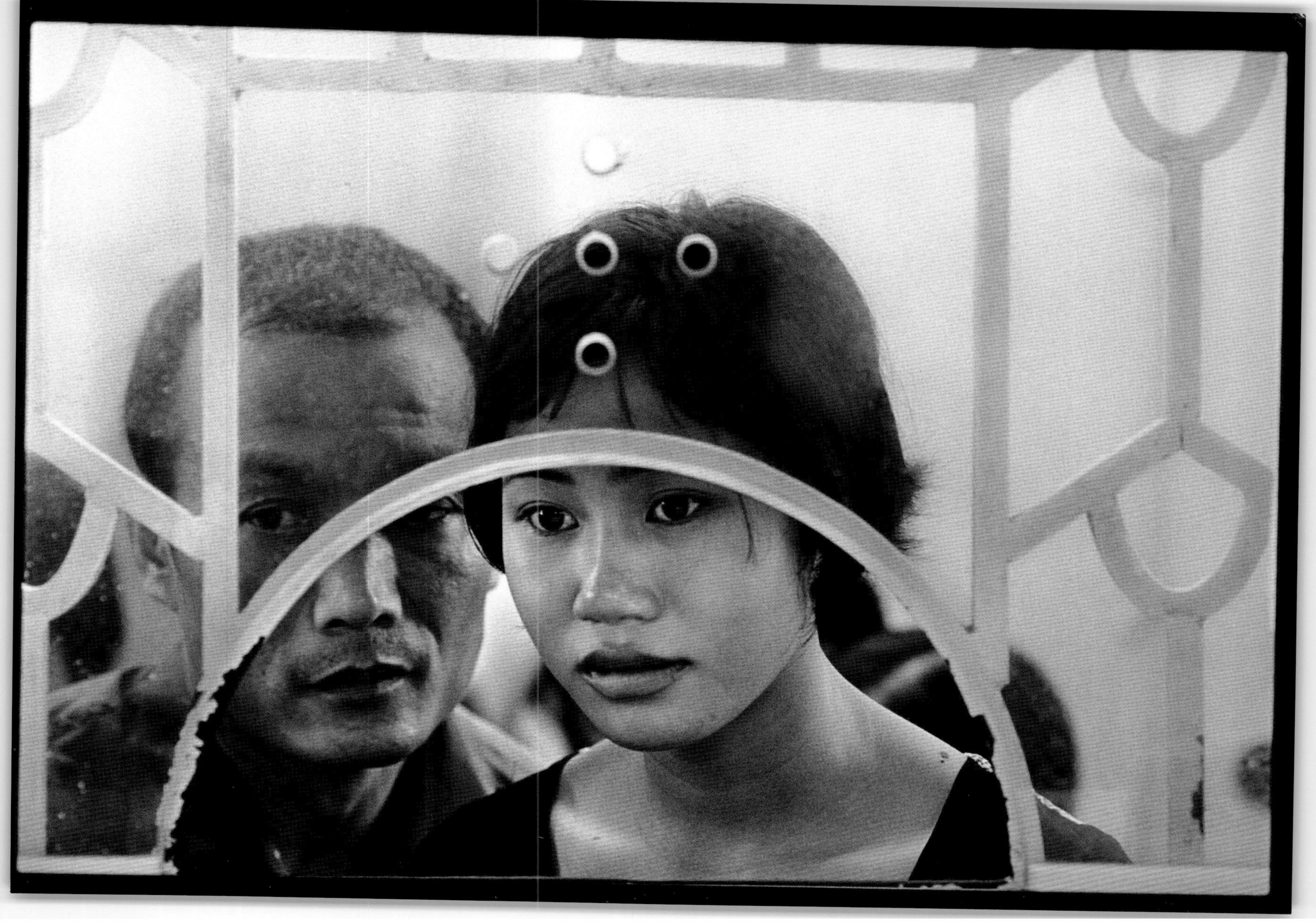

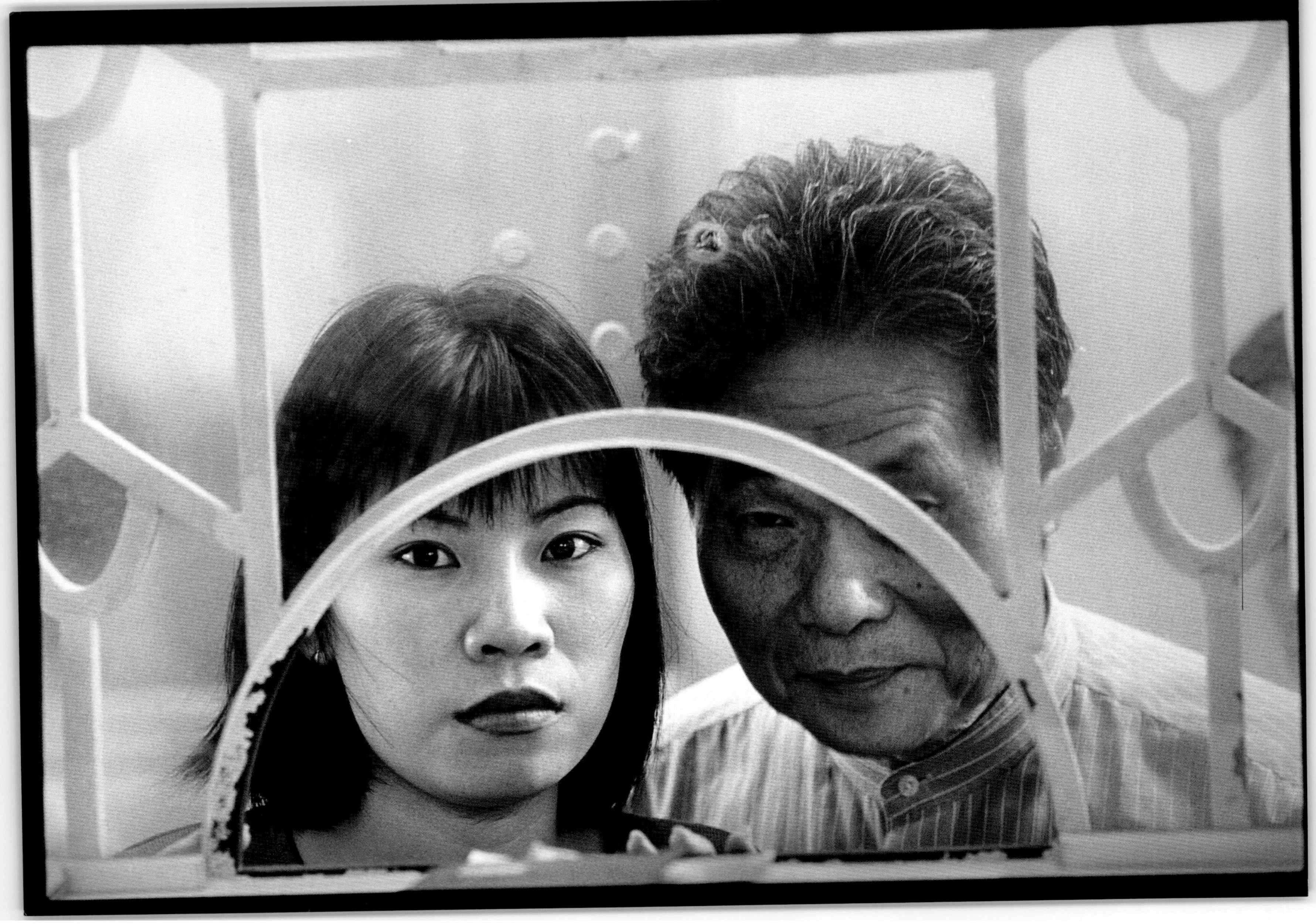

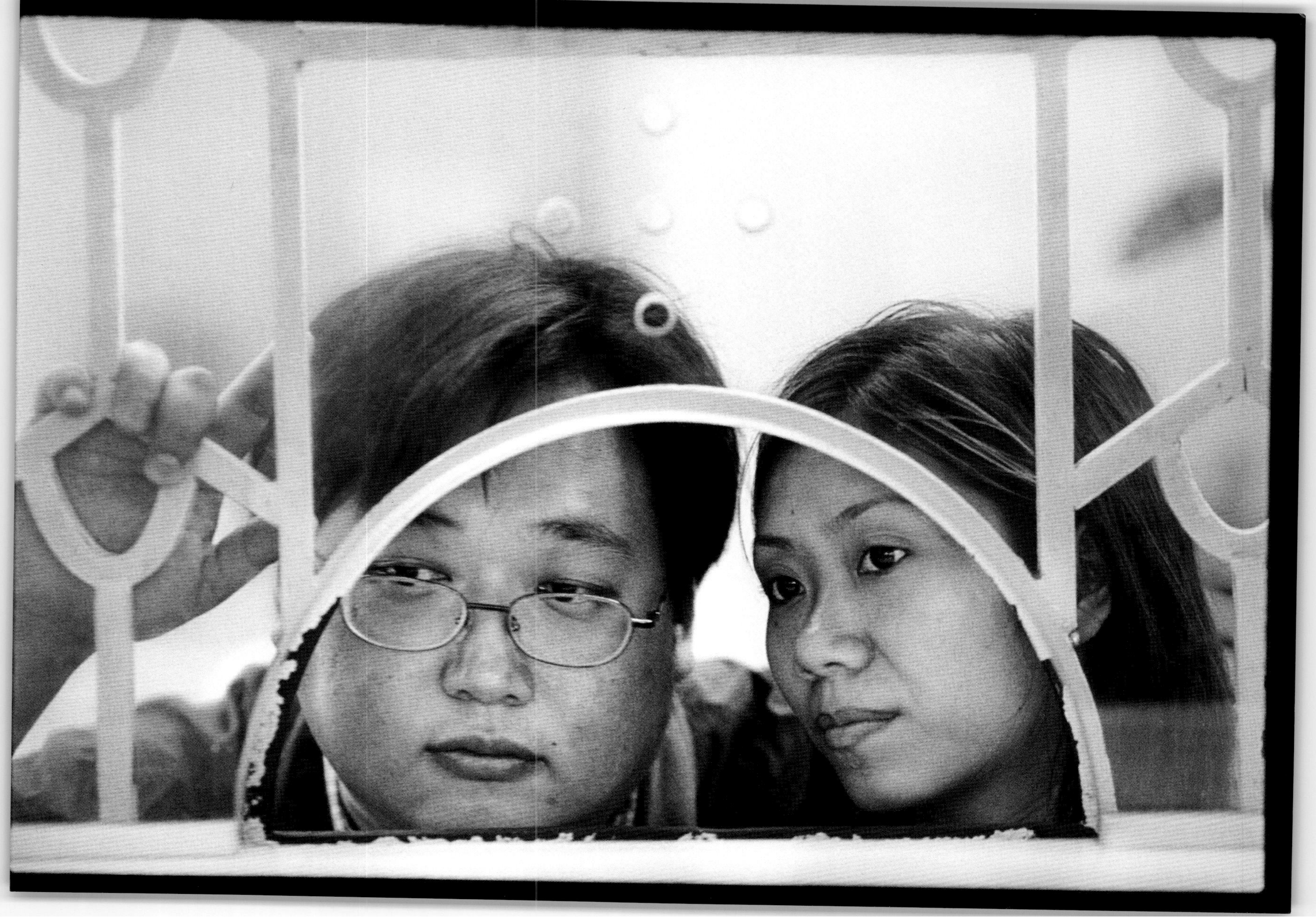

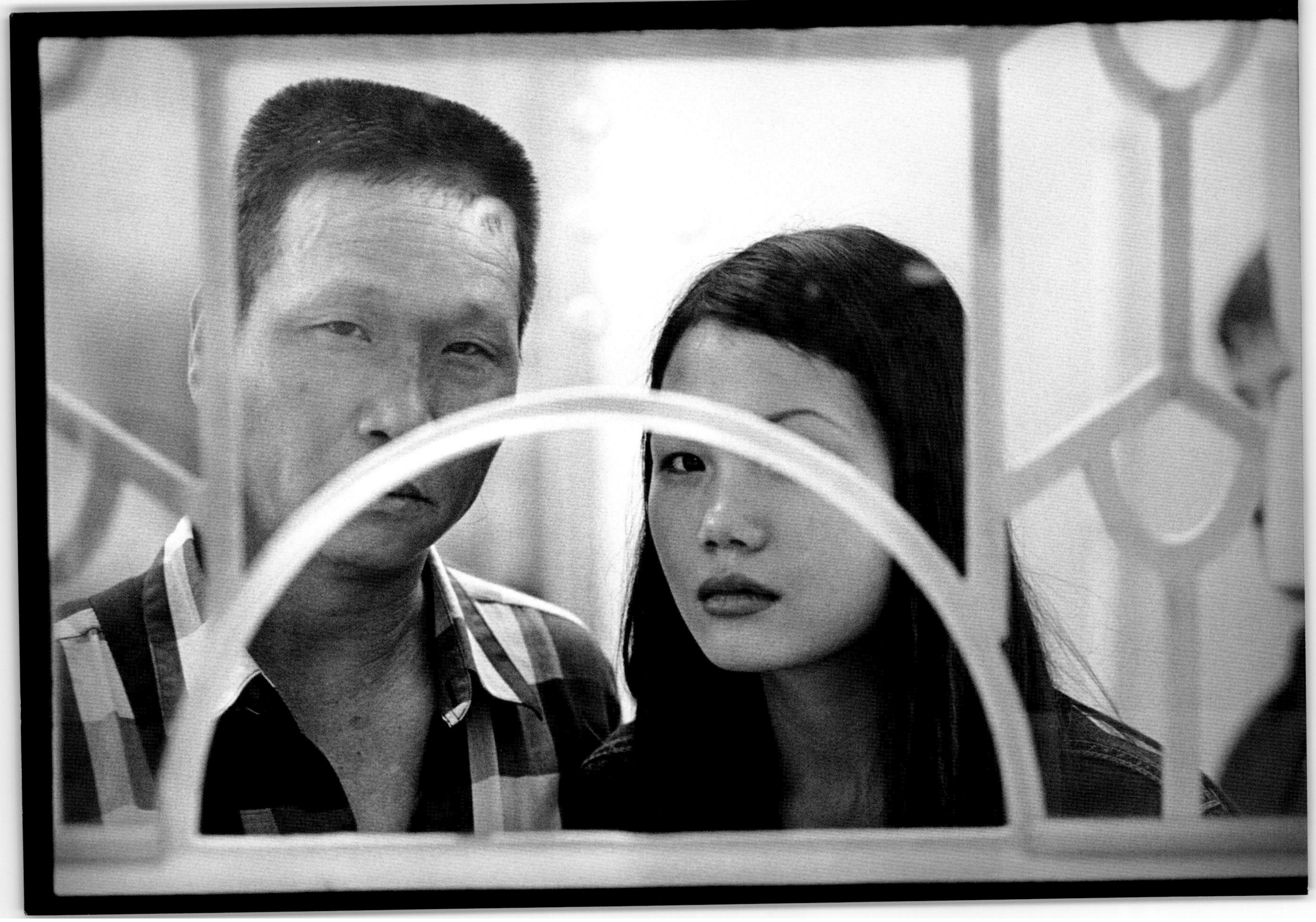

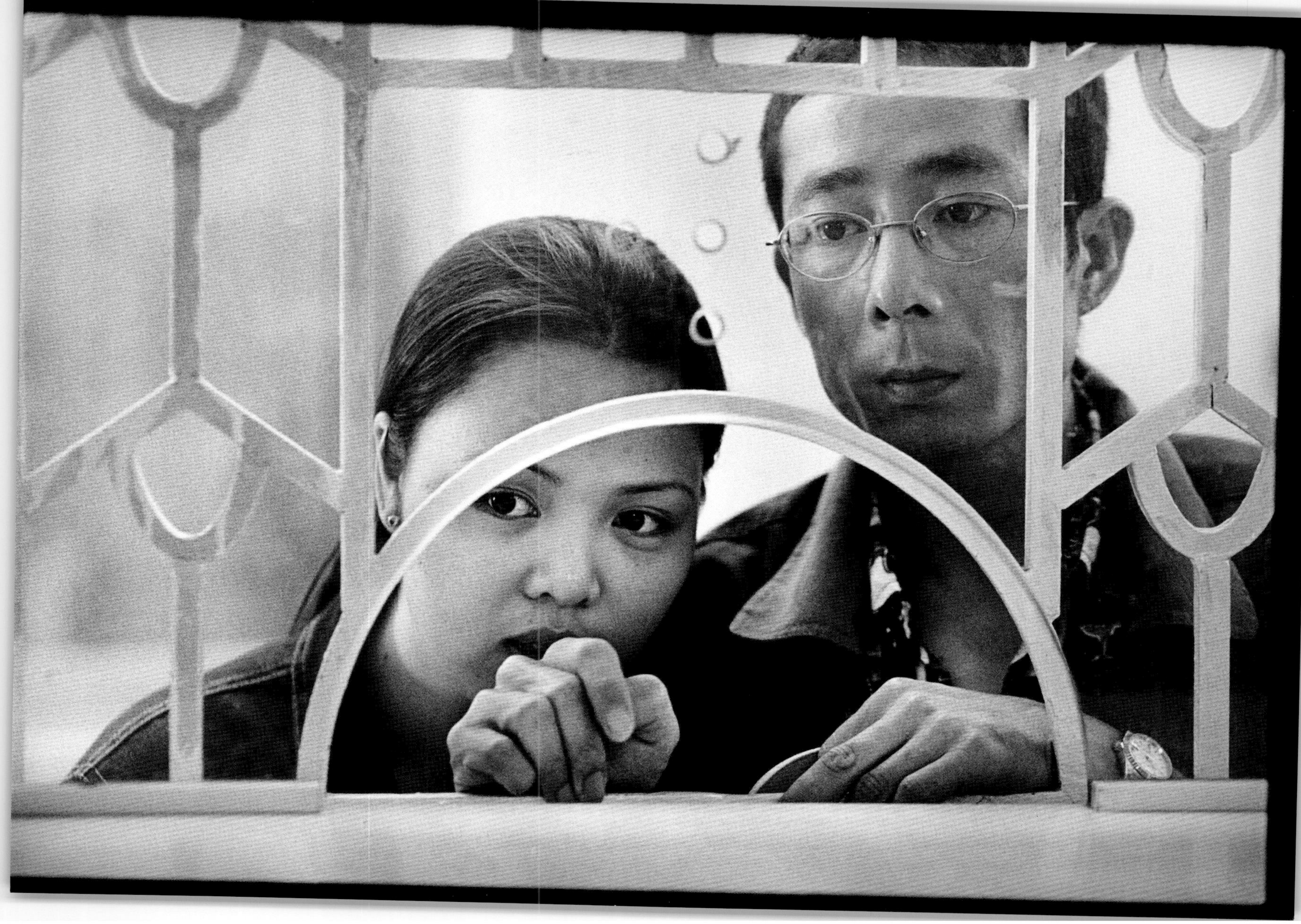

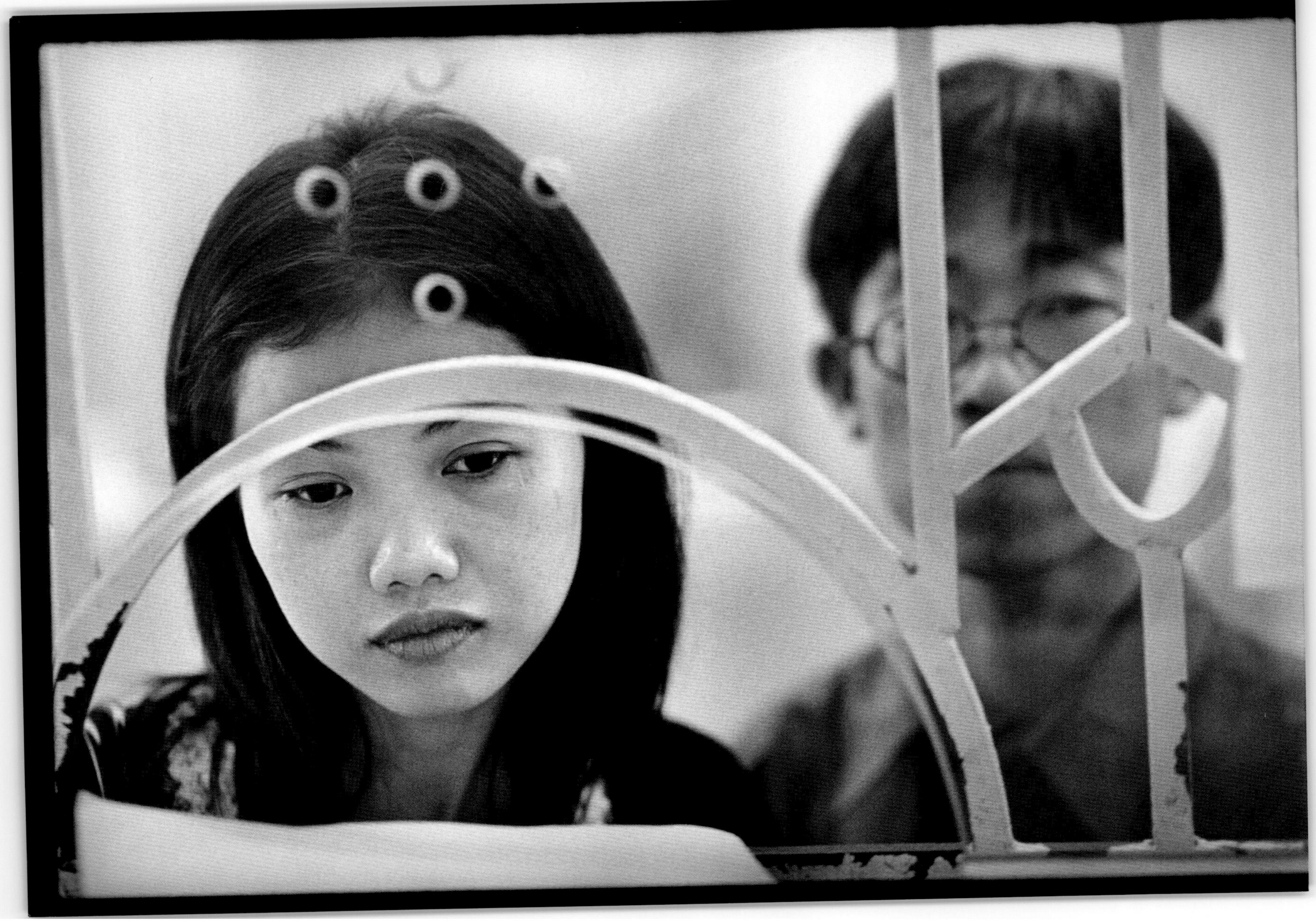

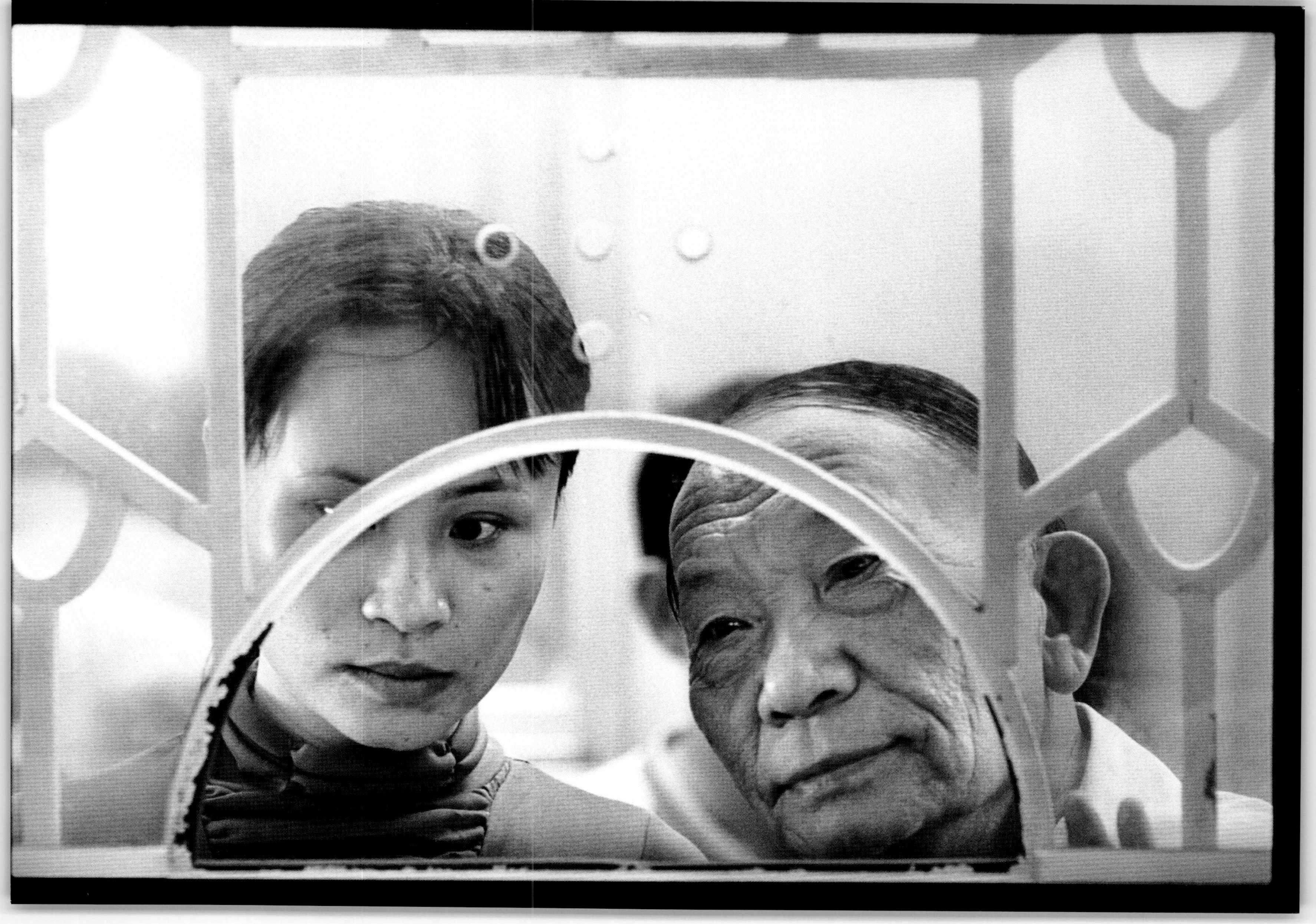

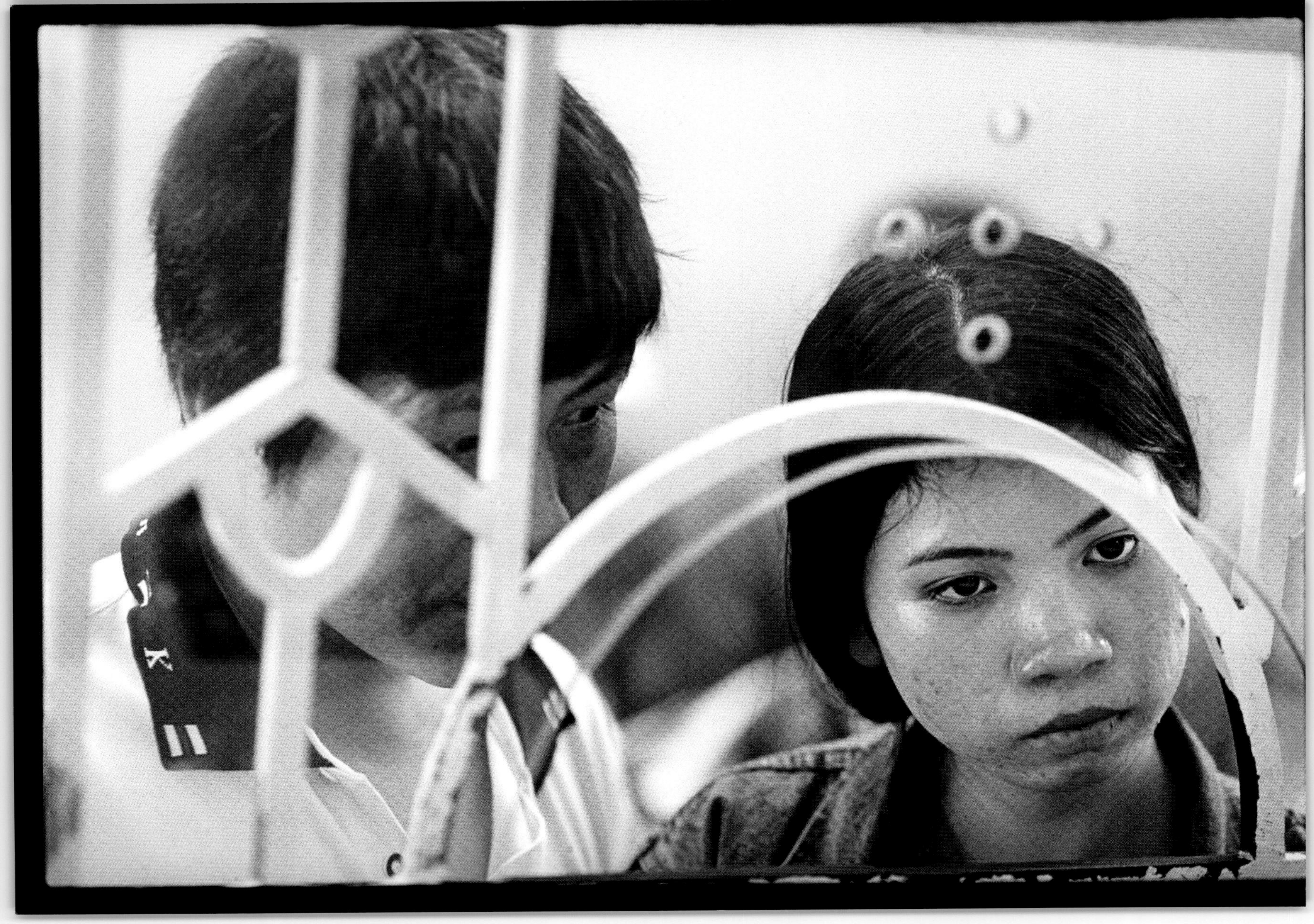

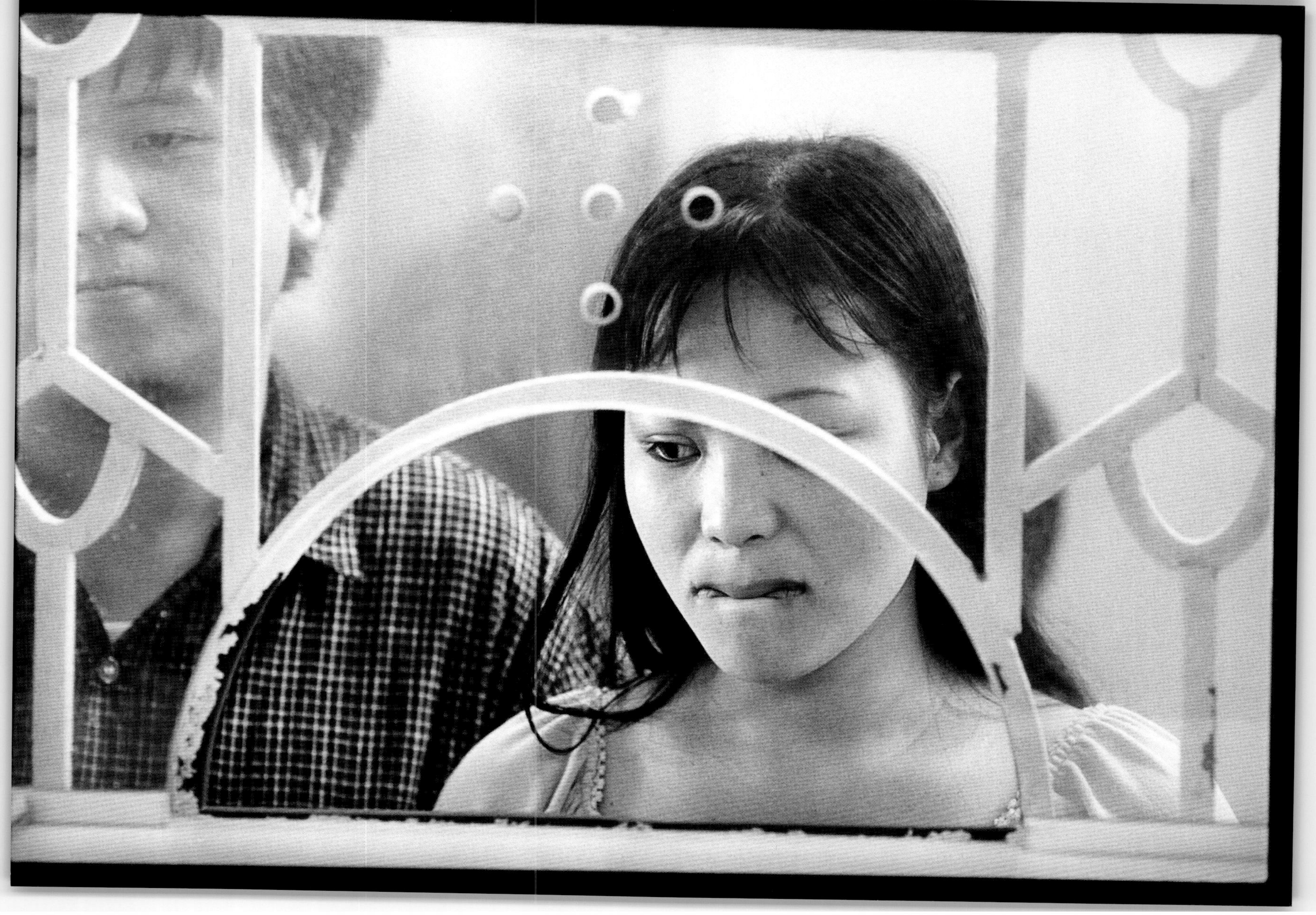

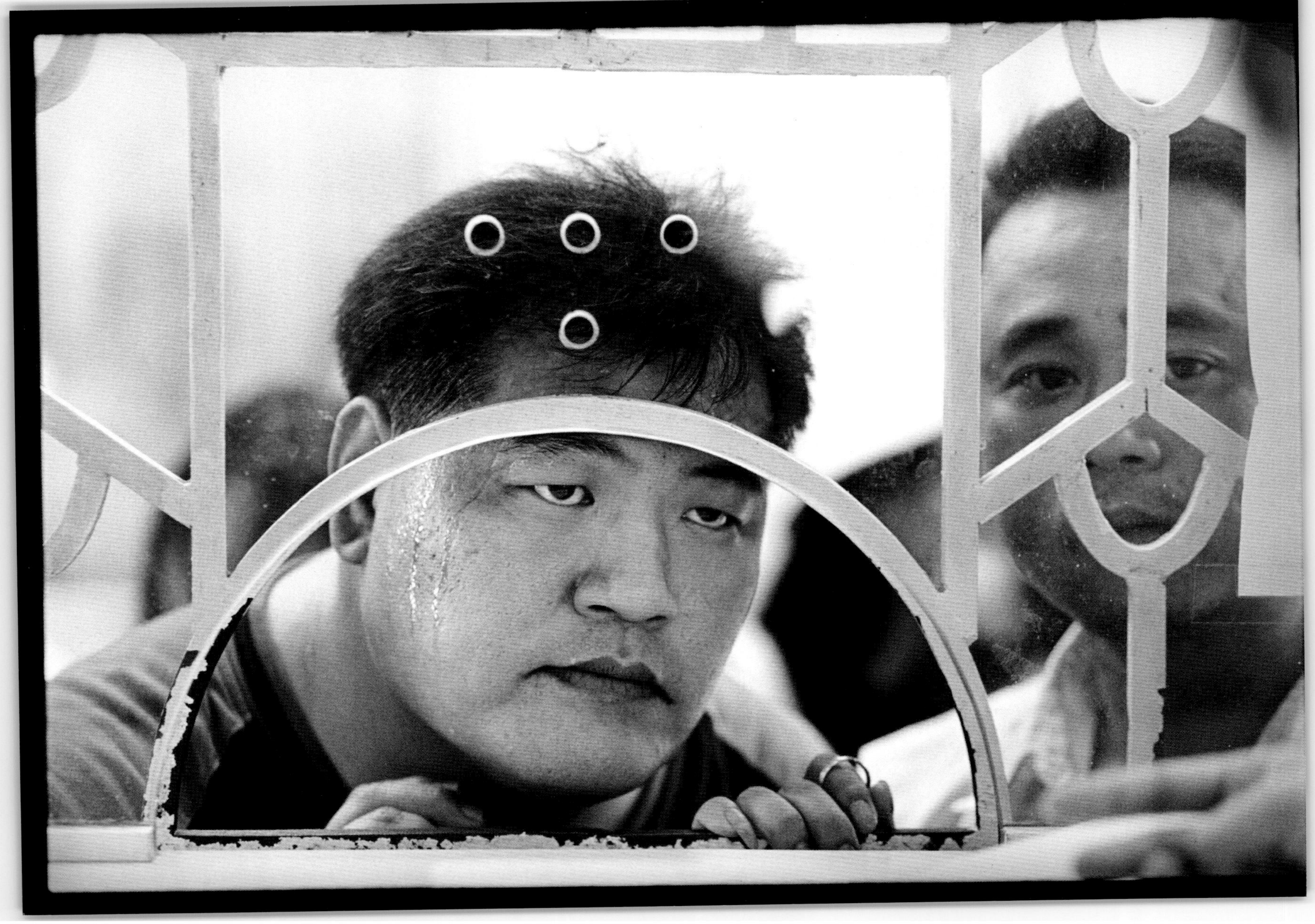

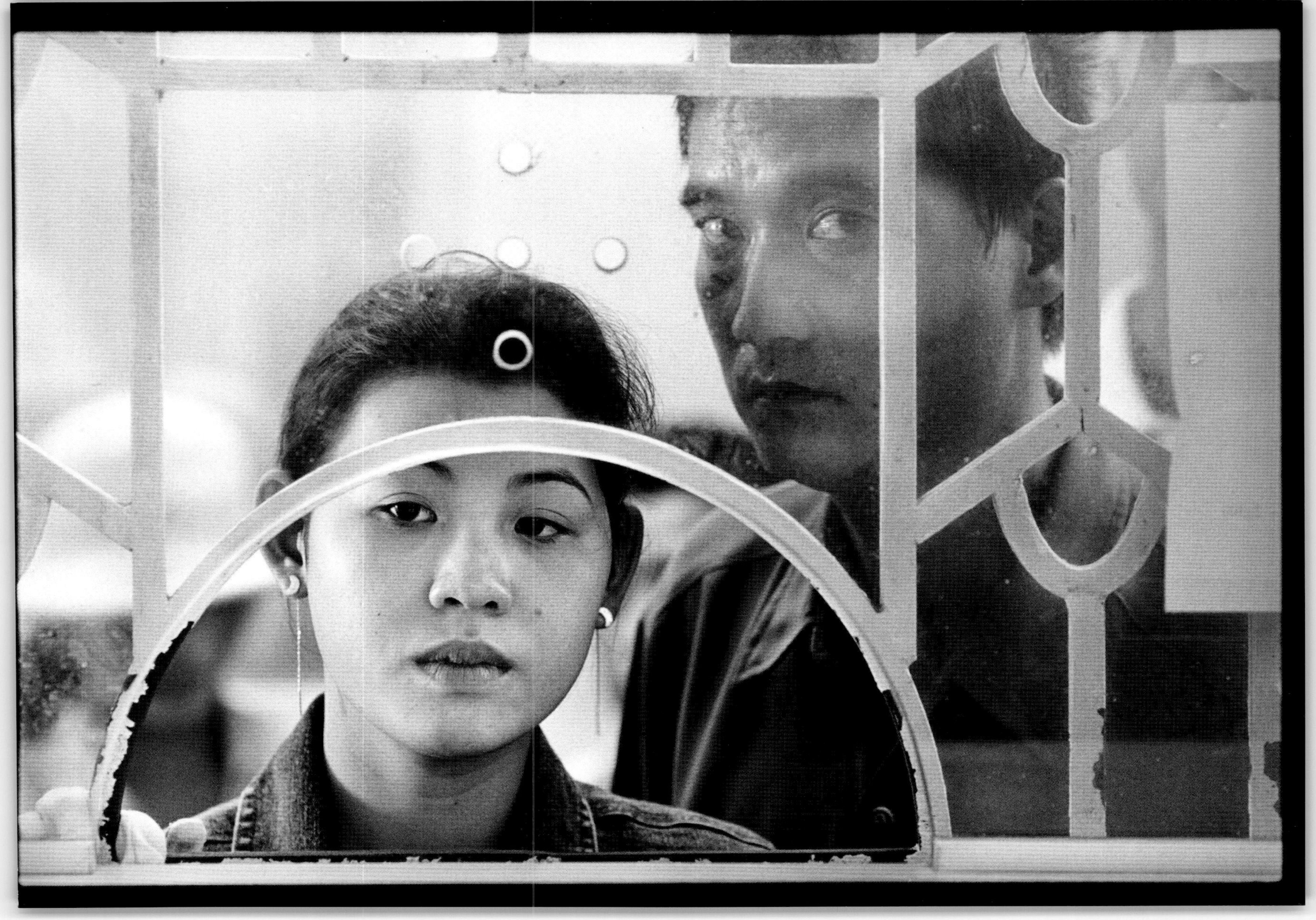

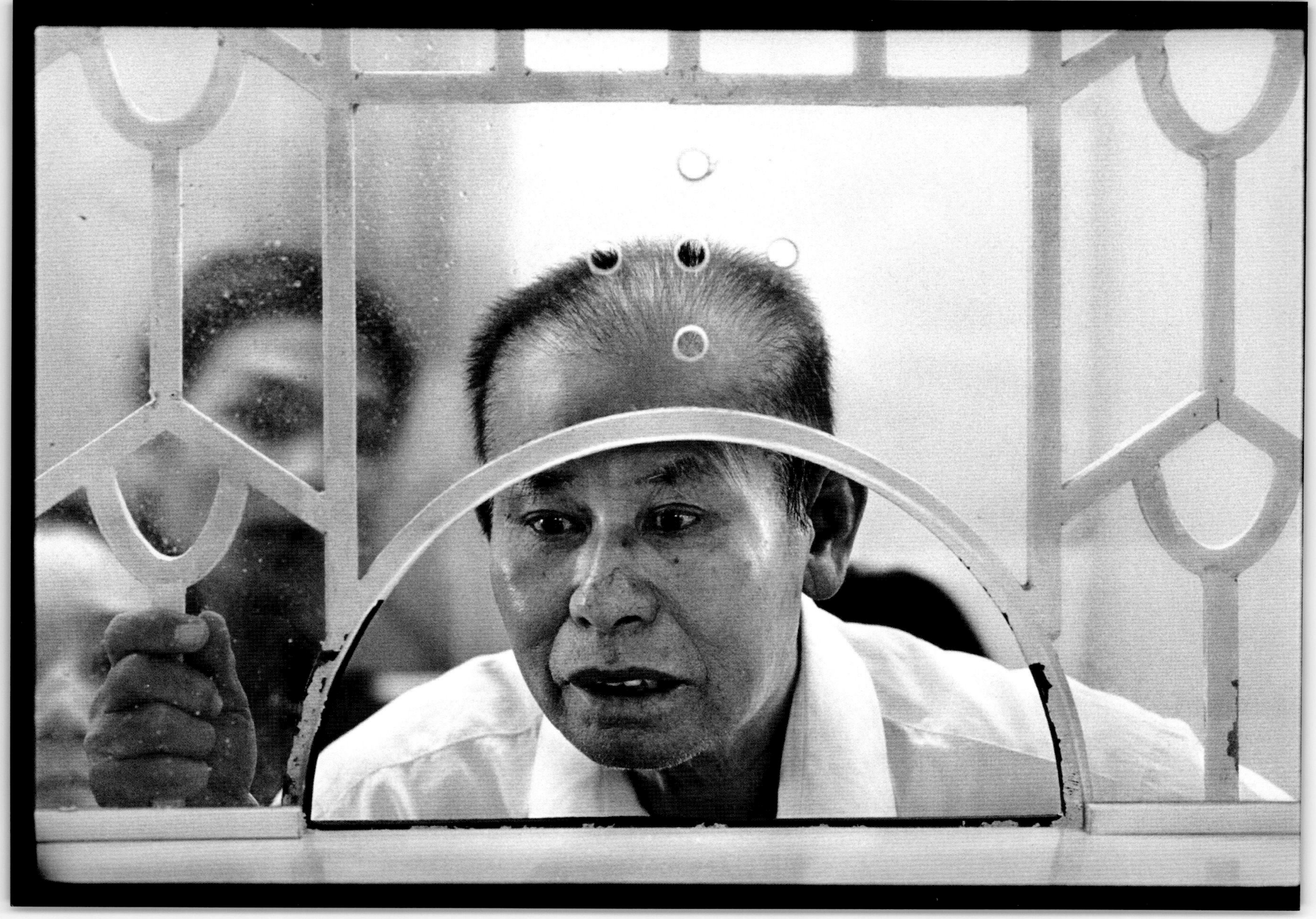

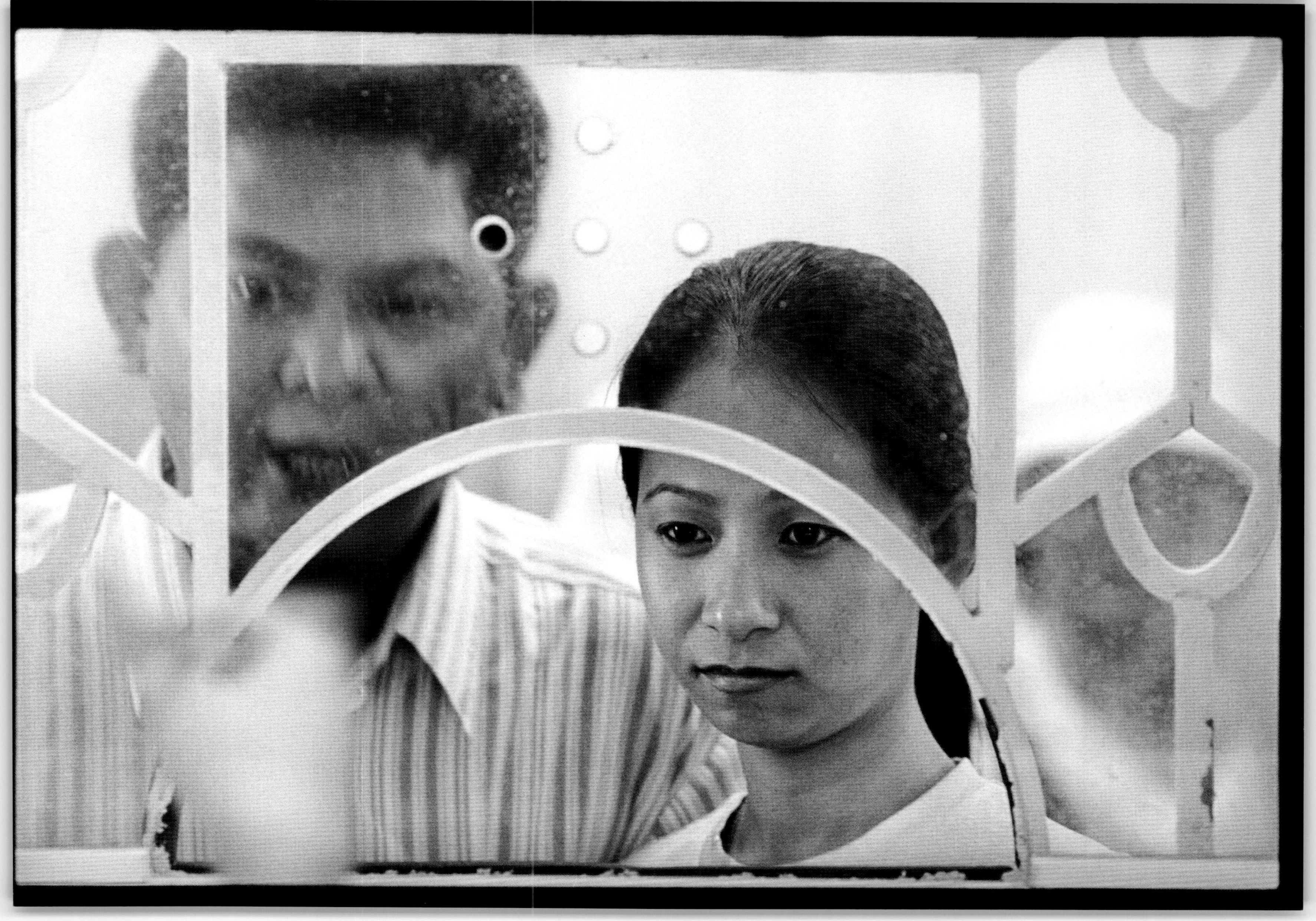

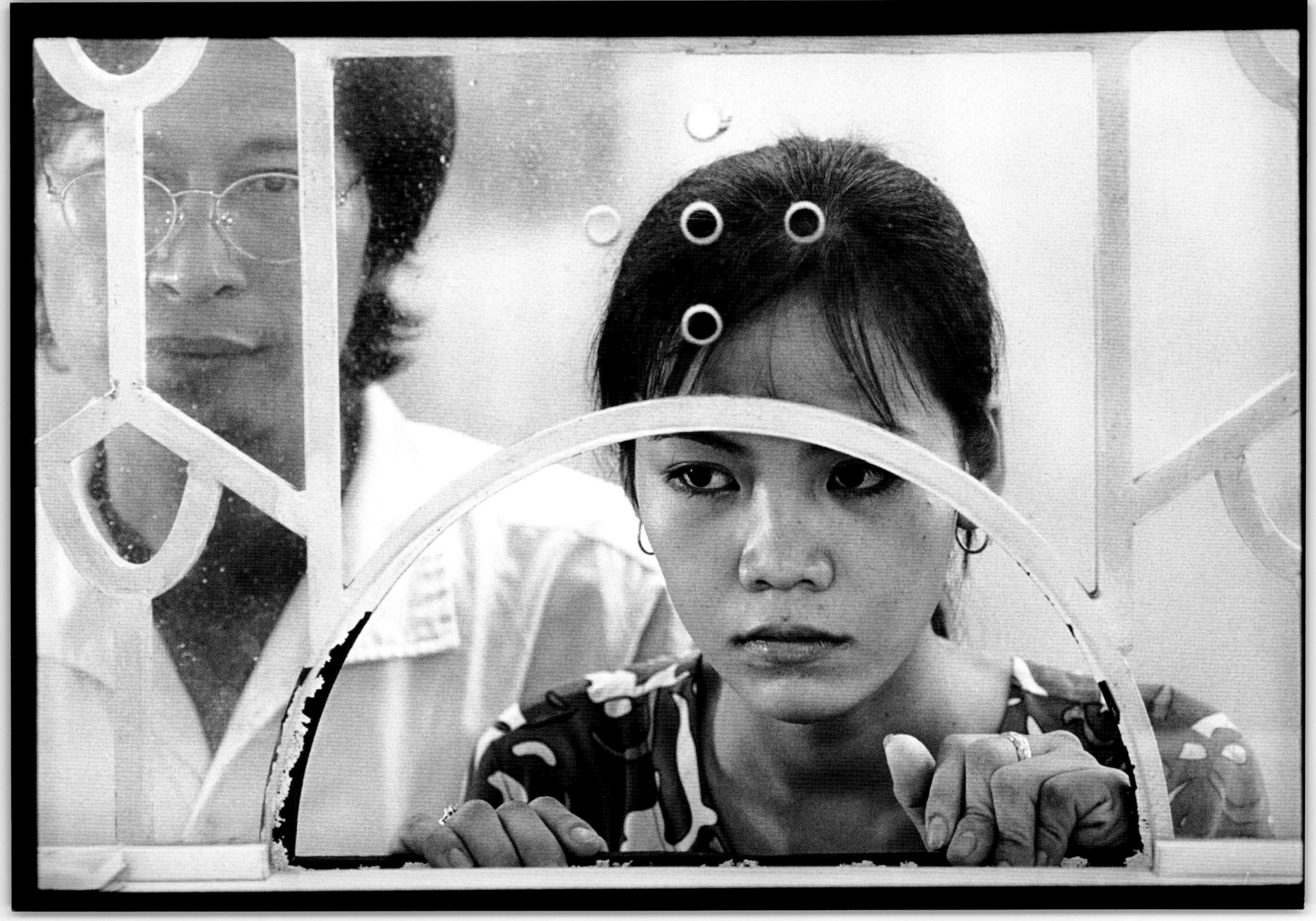

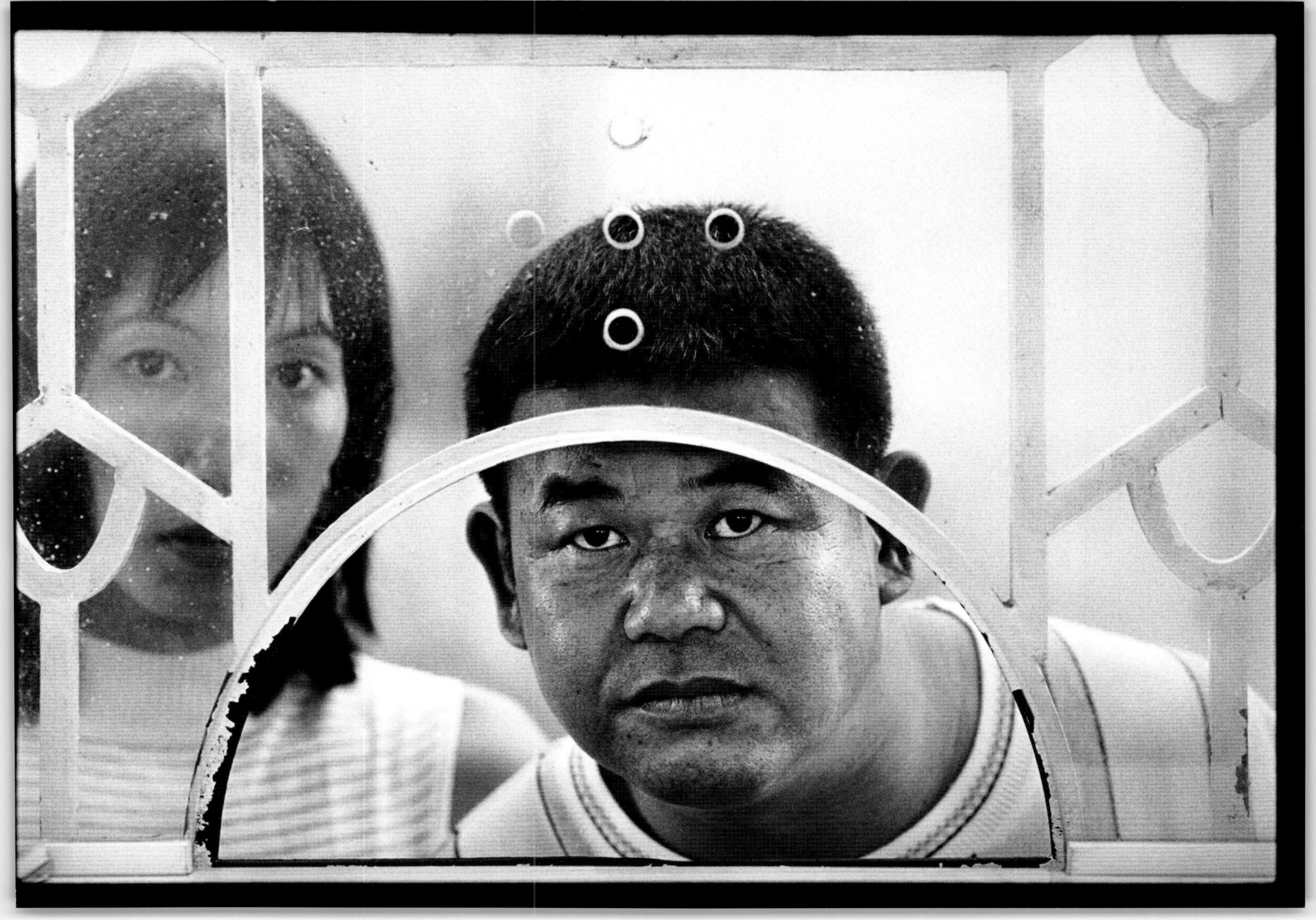

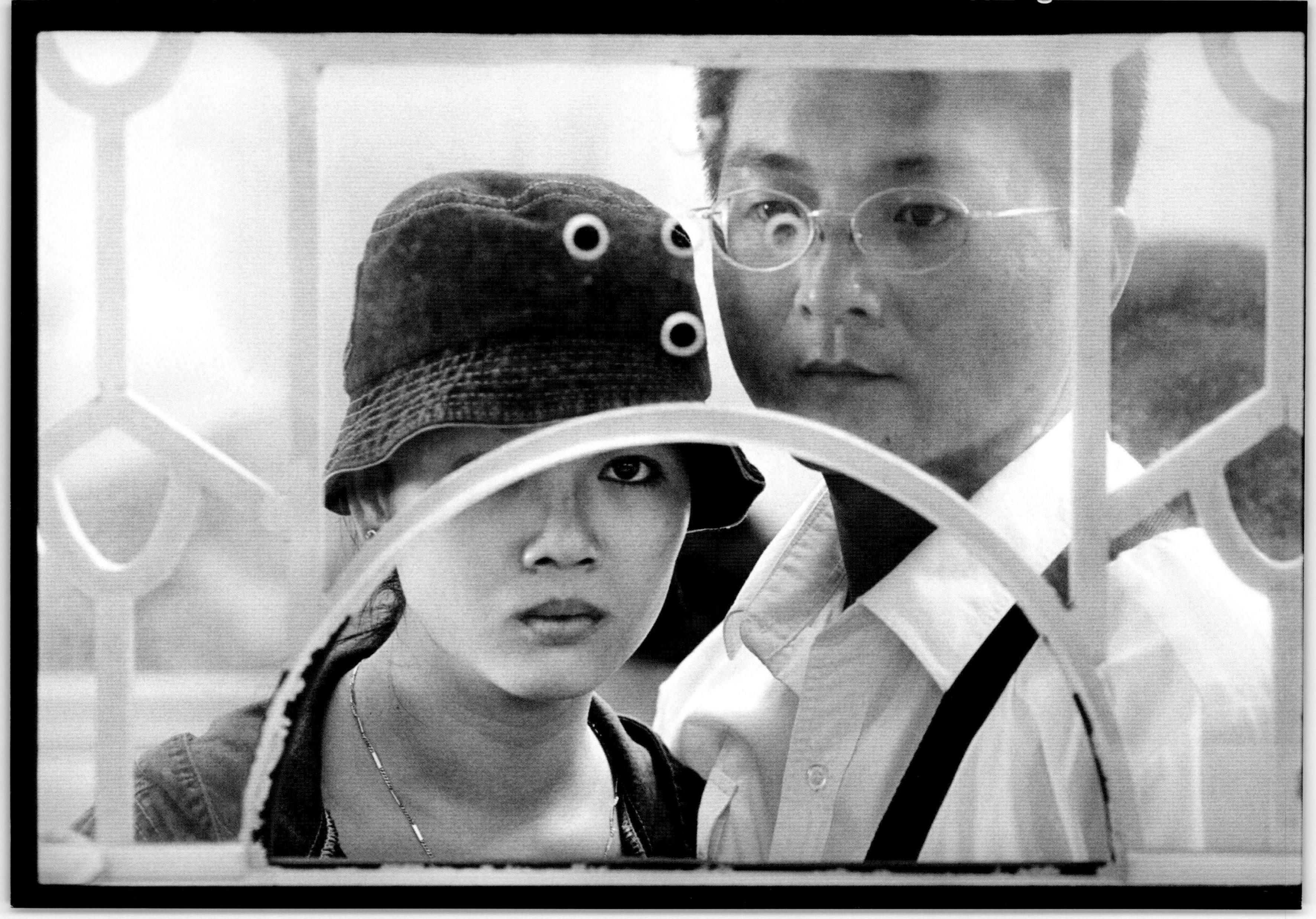

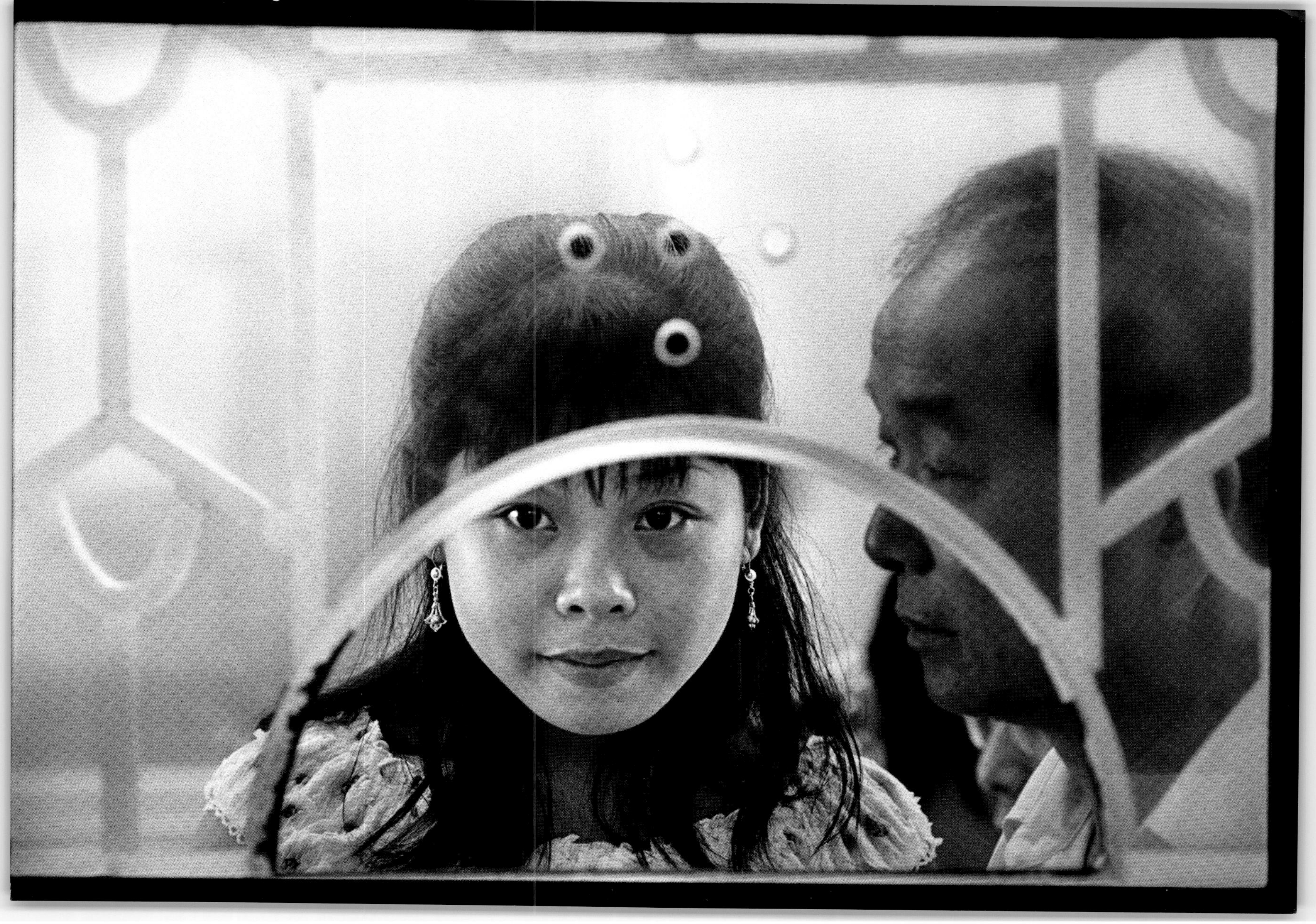

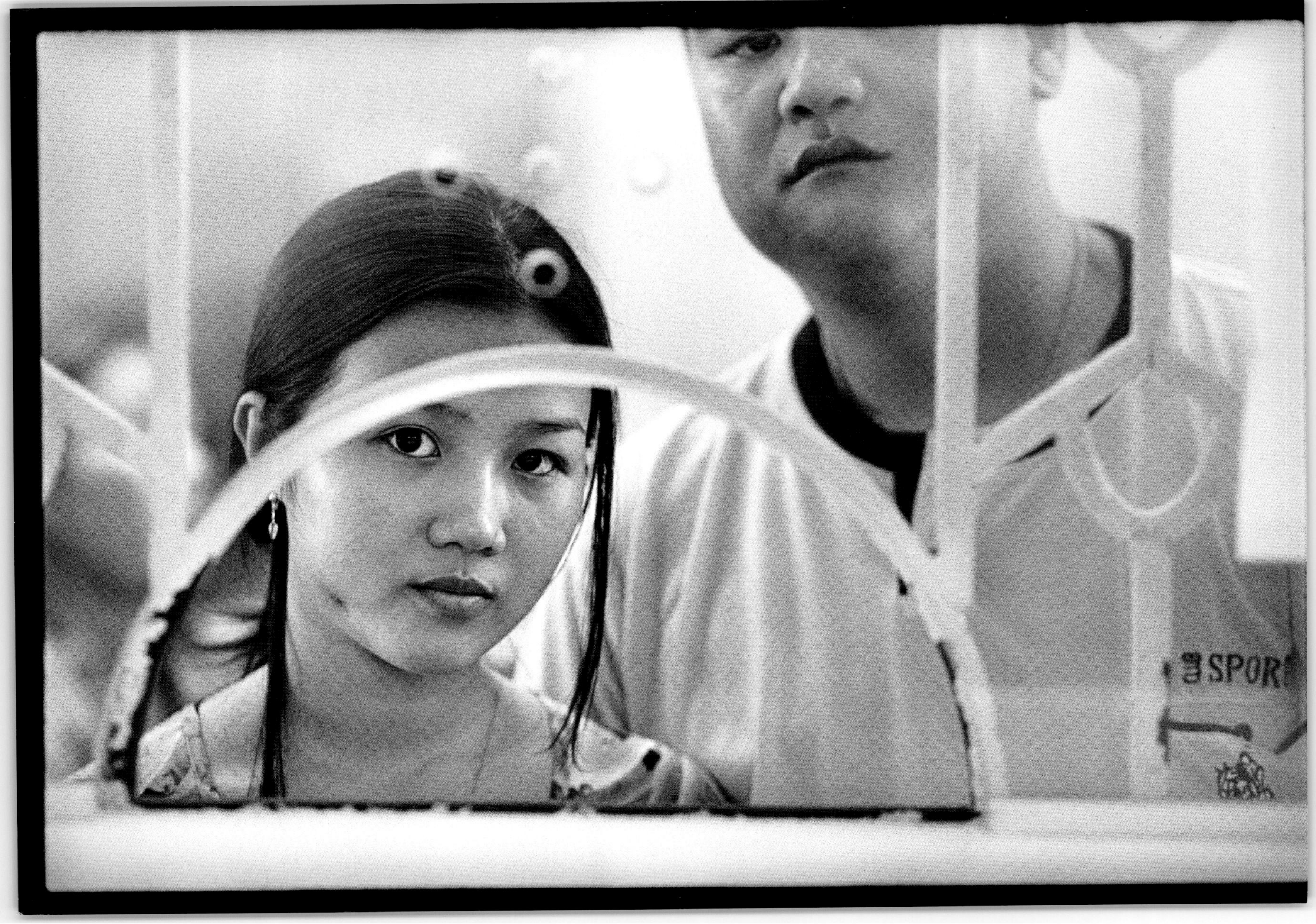

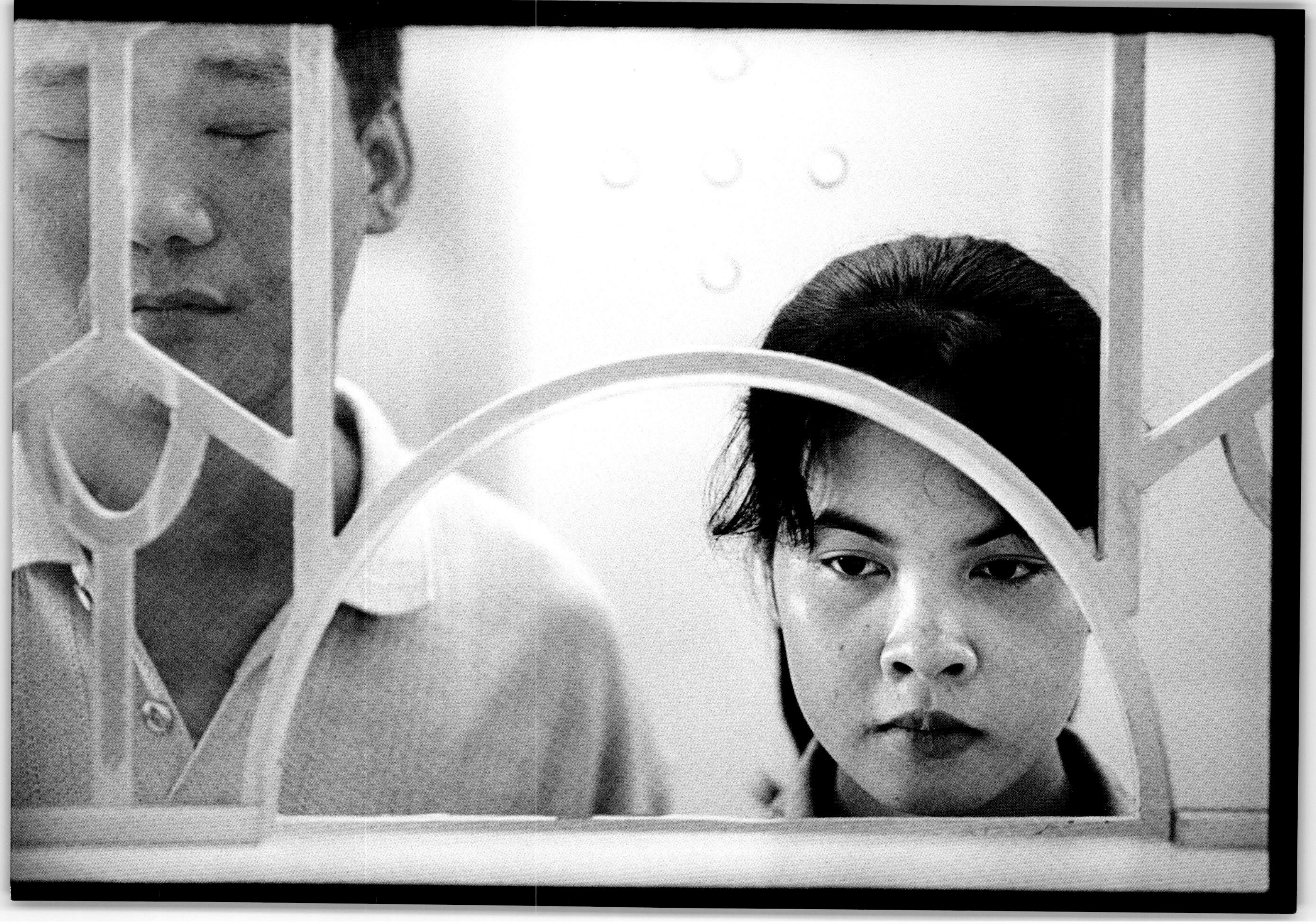

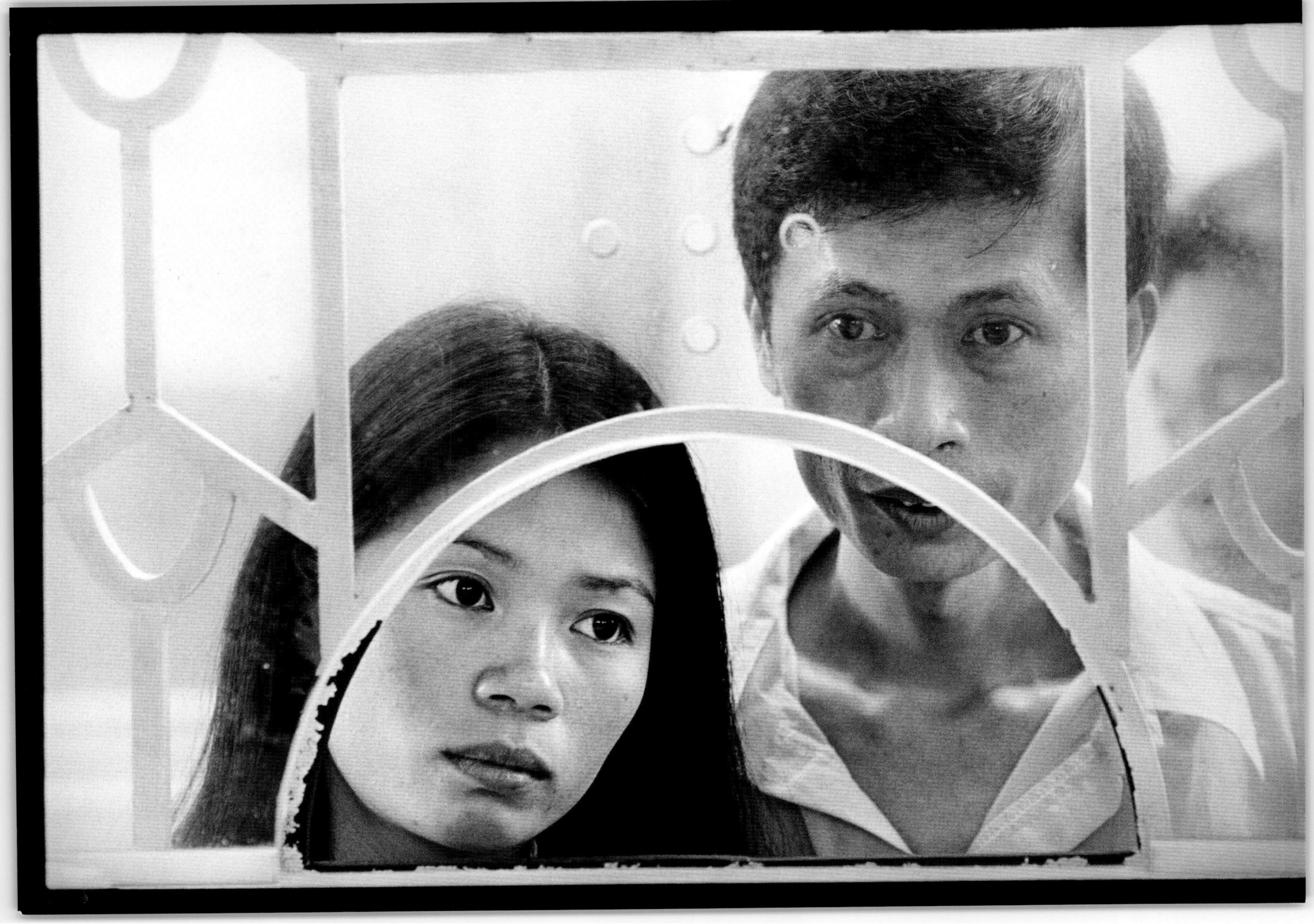

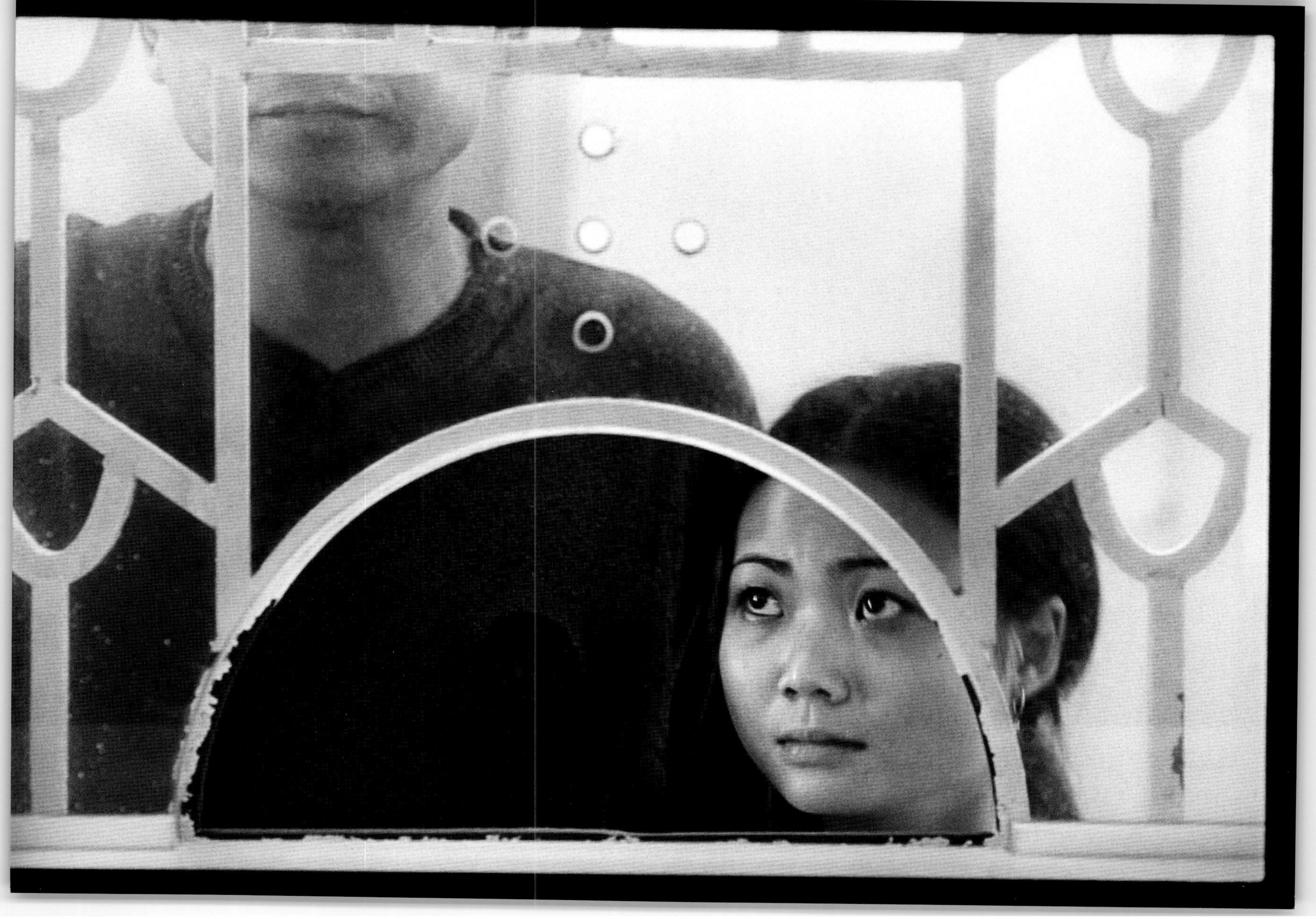

Before visas can be processed, the couples are counseled by Taiwanese officials about the problems of cross-cultural adjustment.

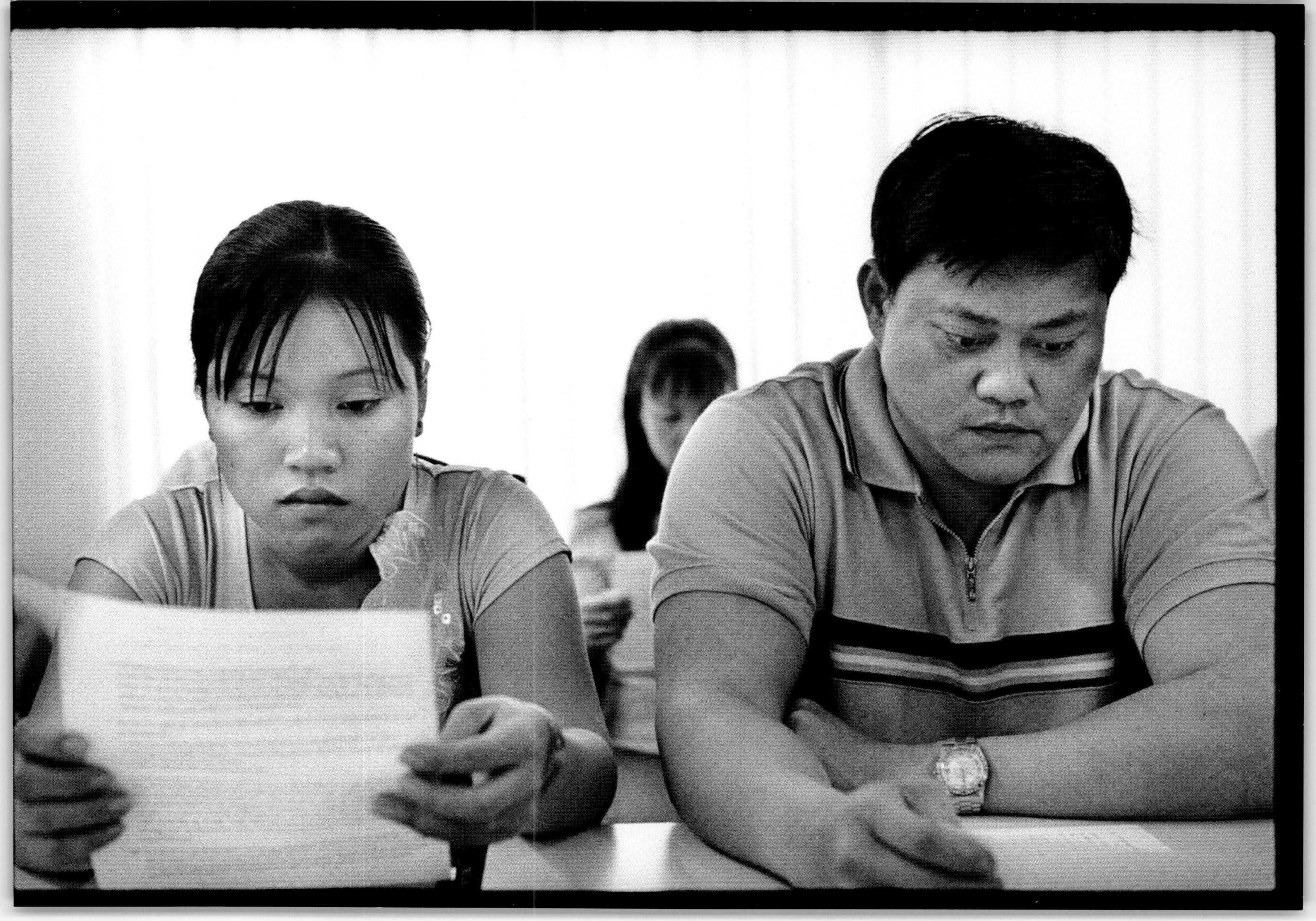

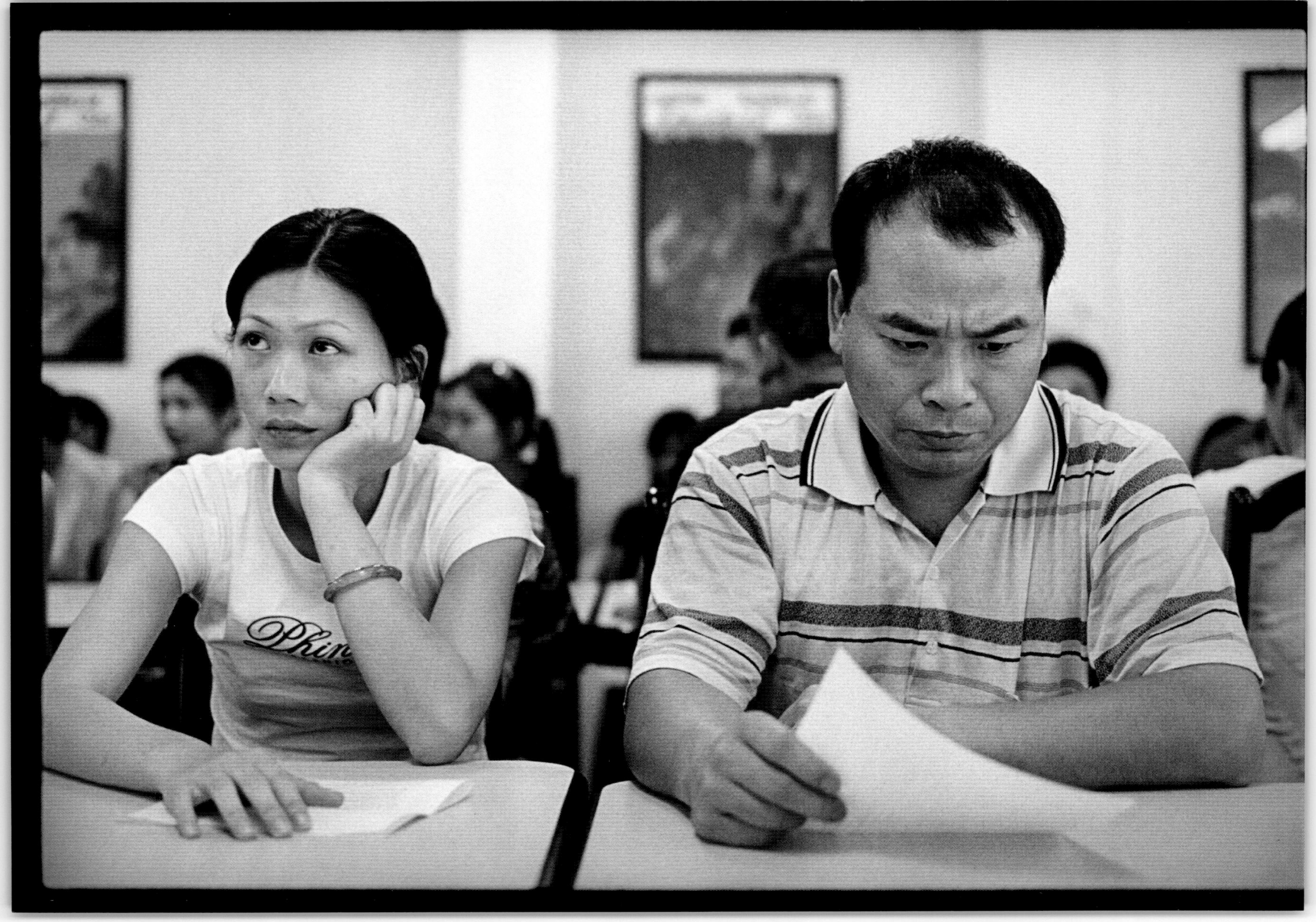

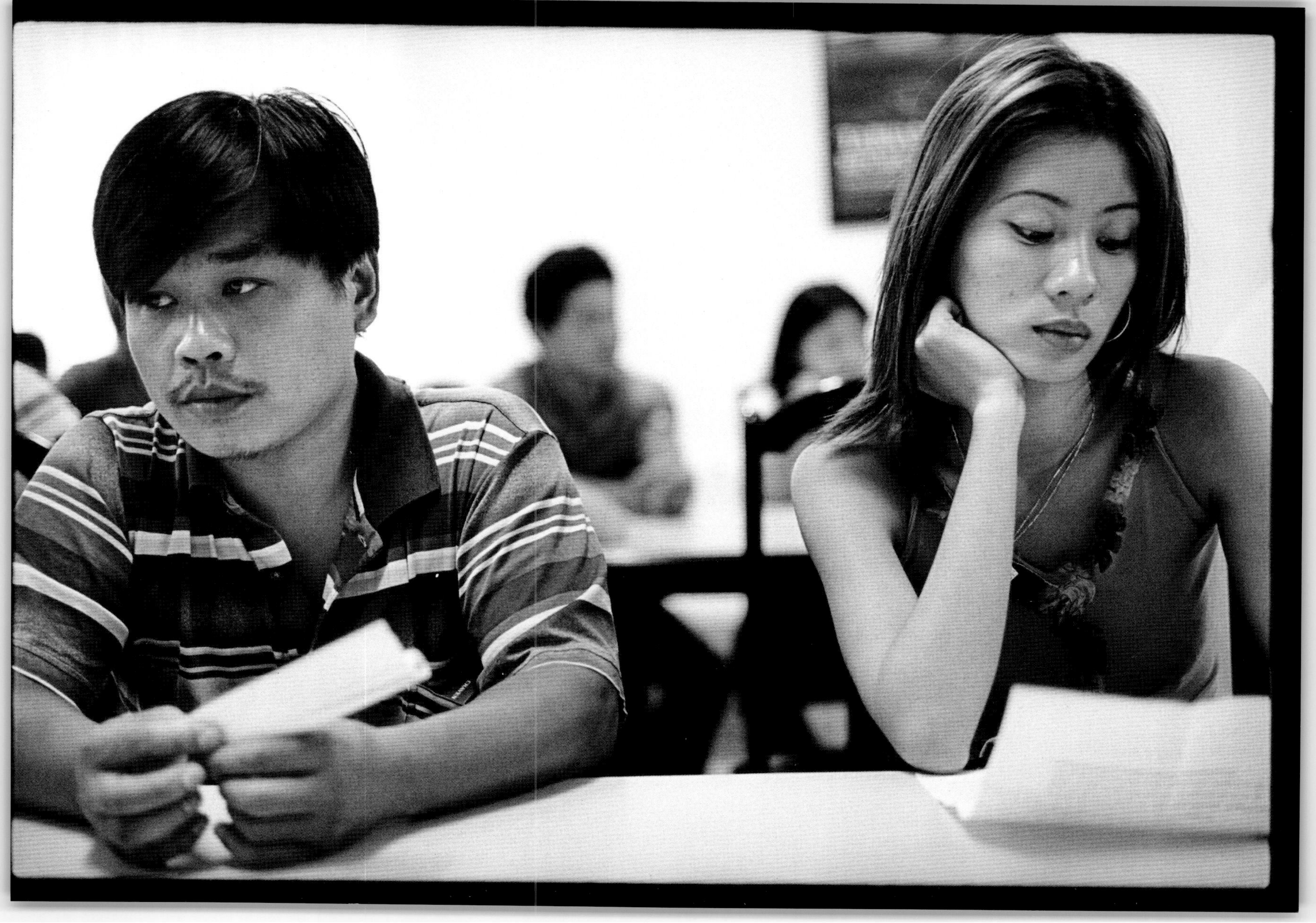

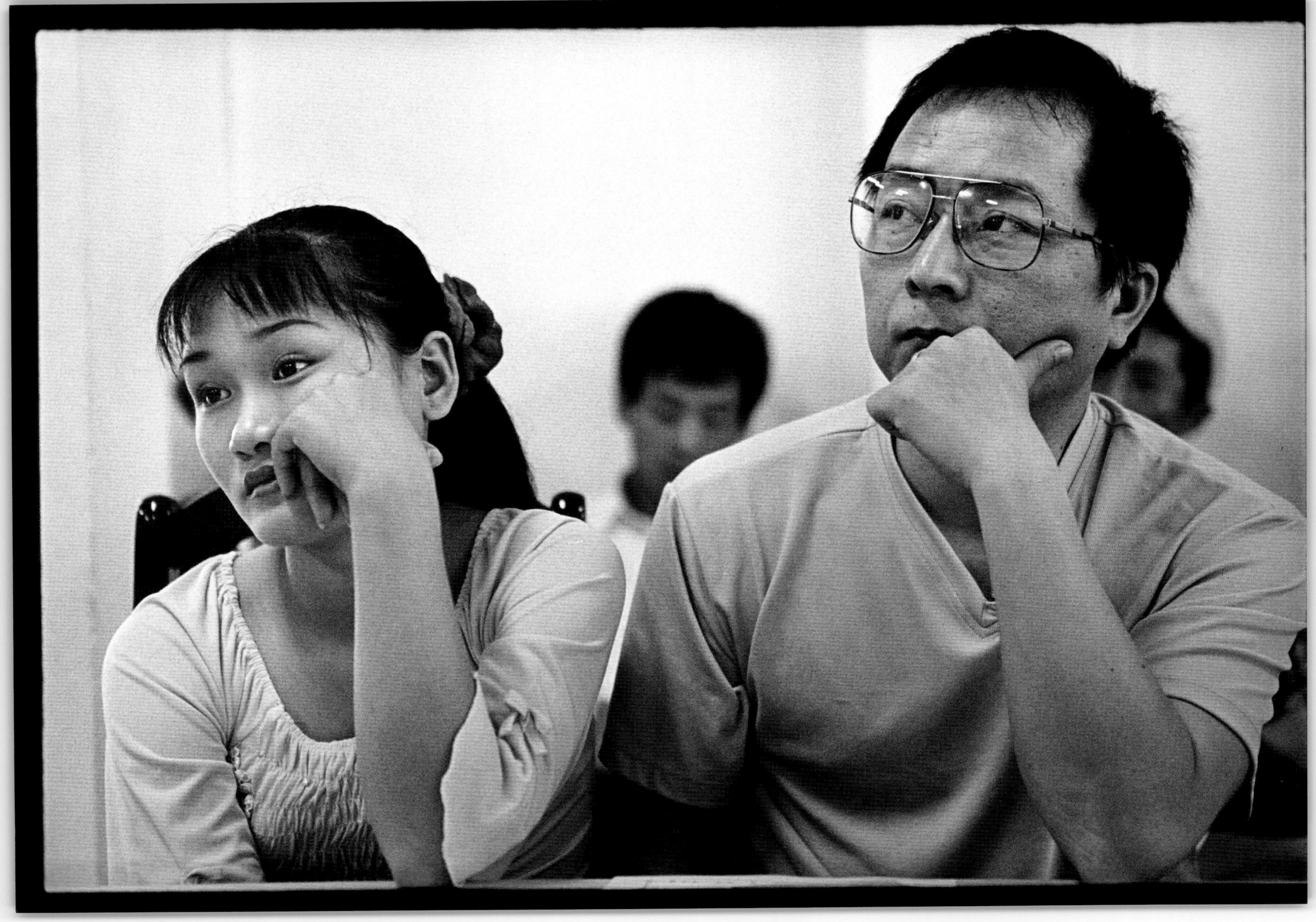

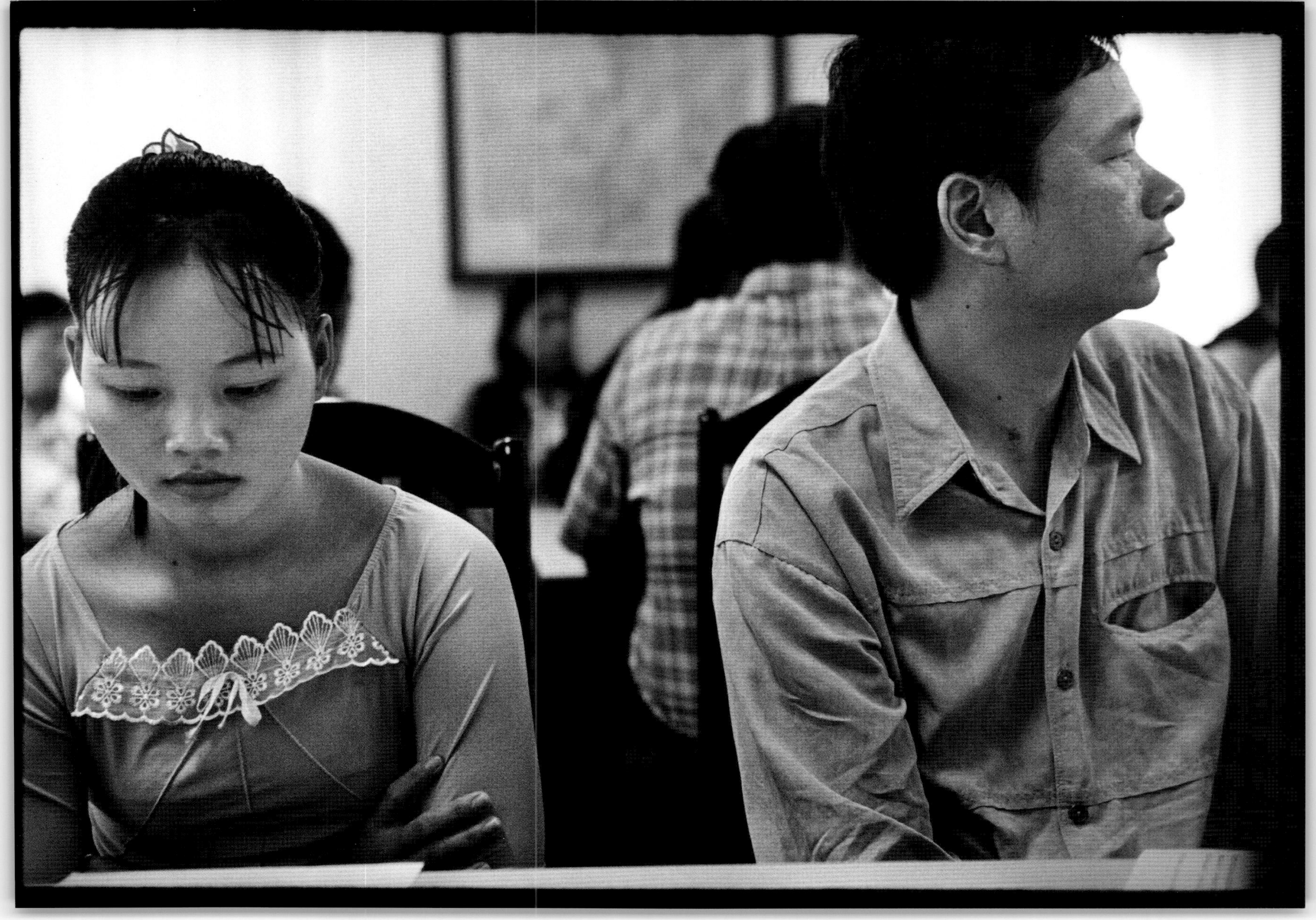

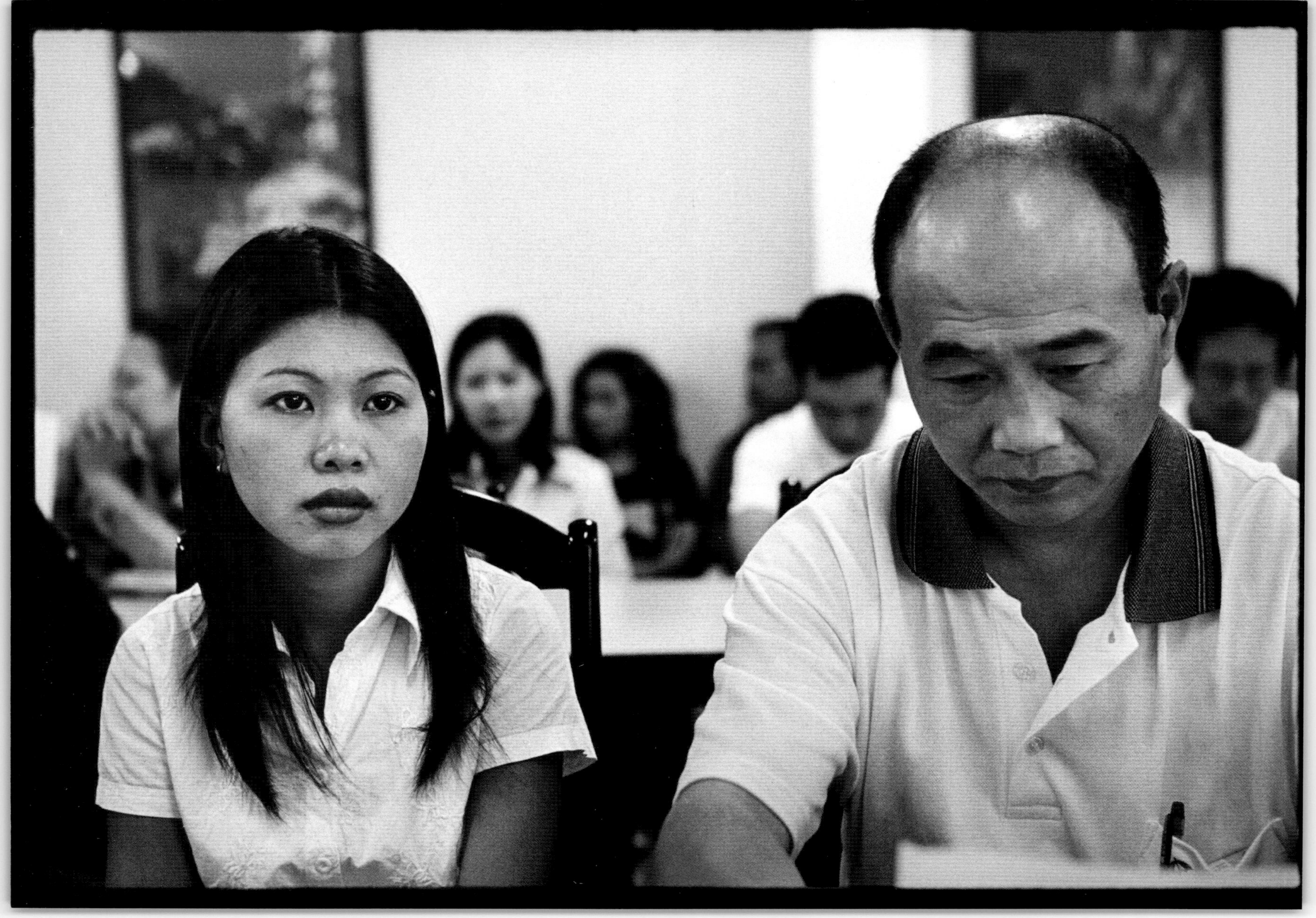

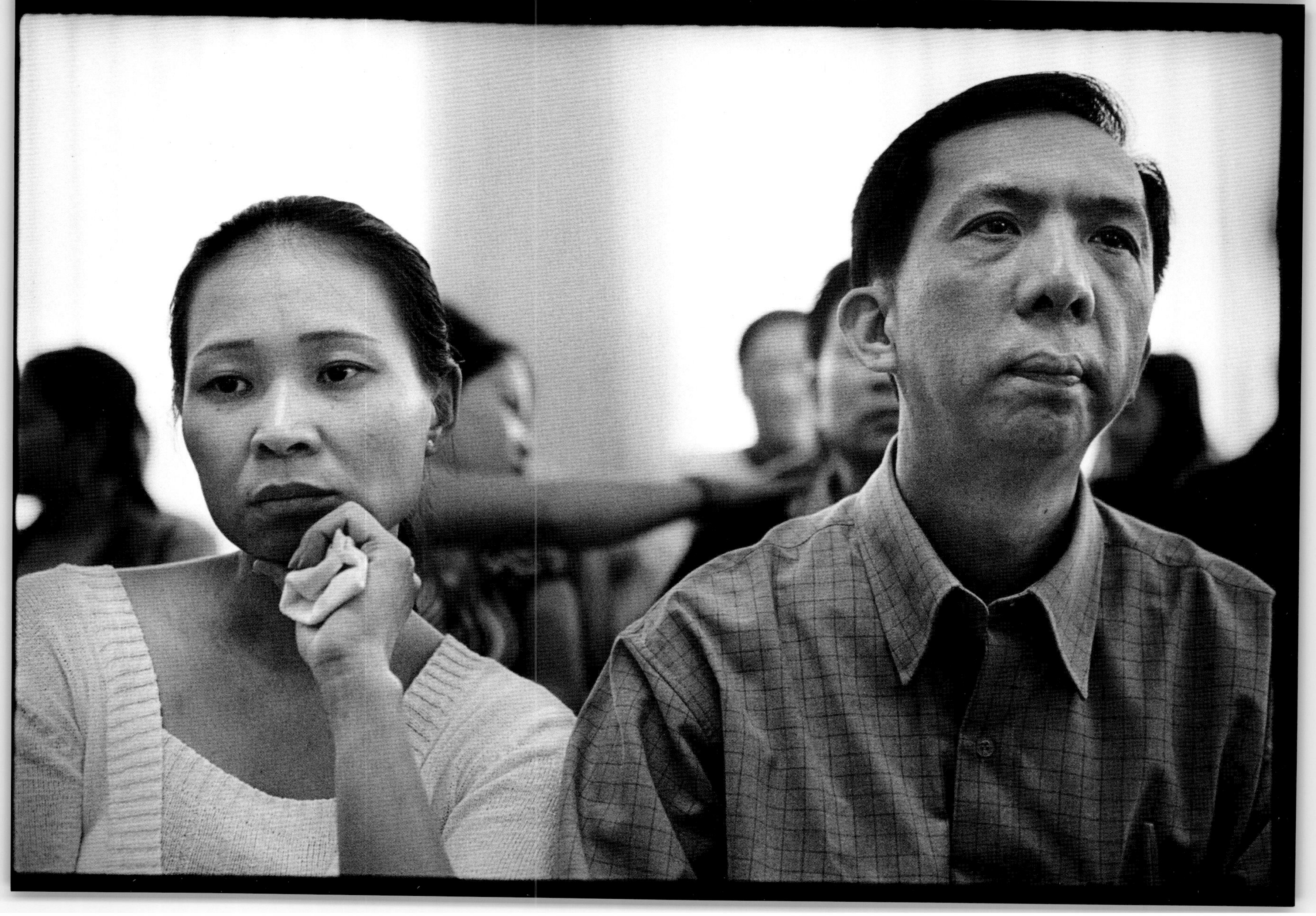

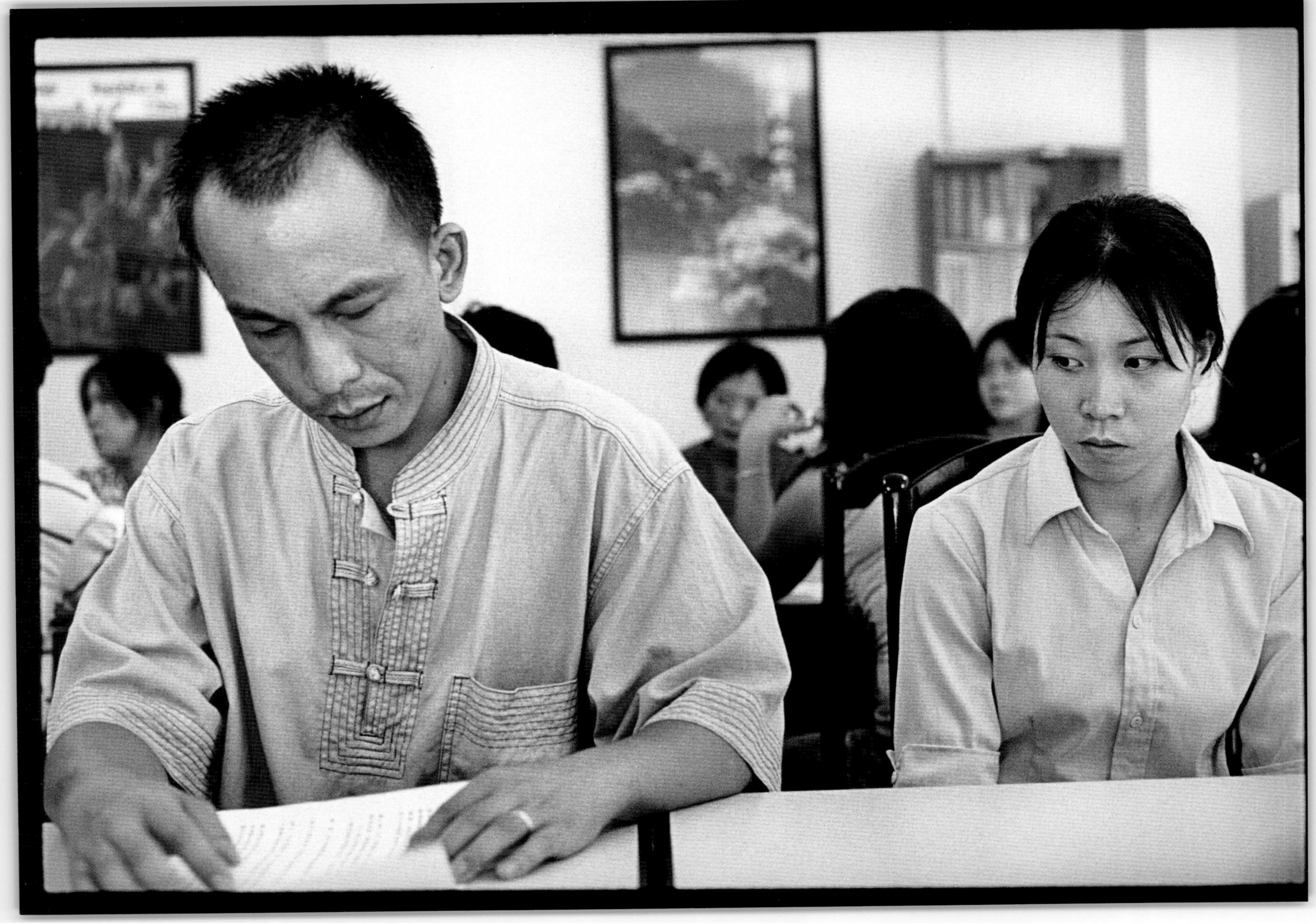

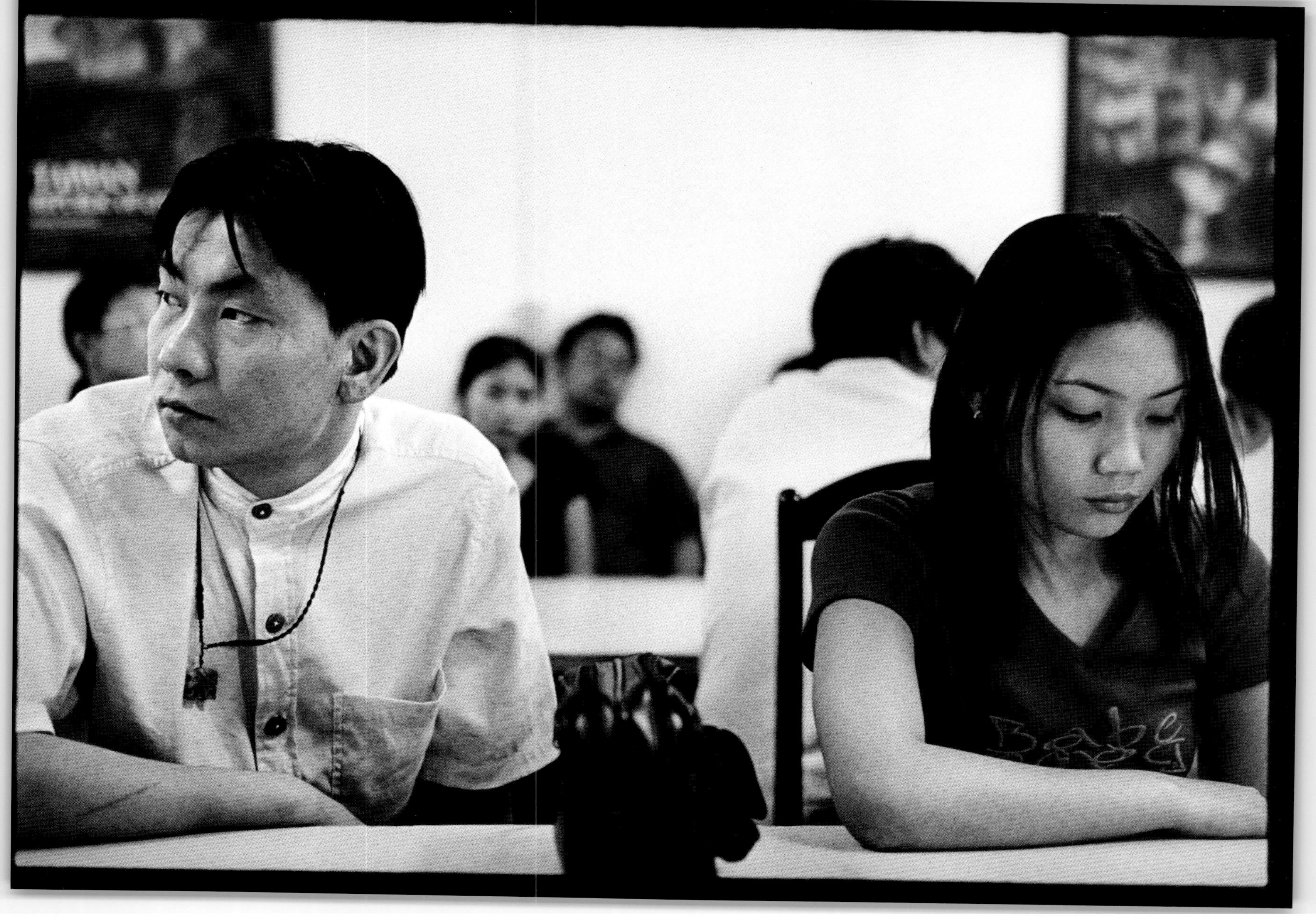

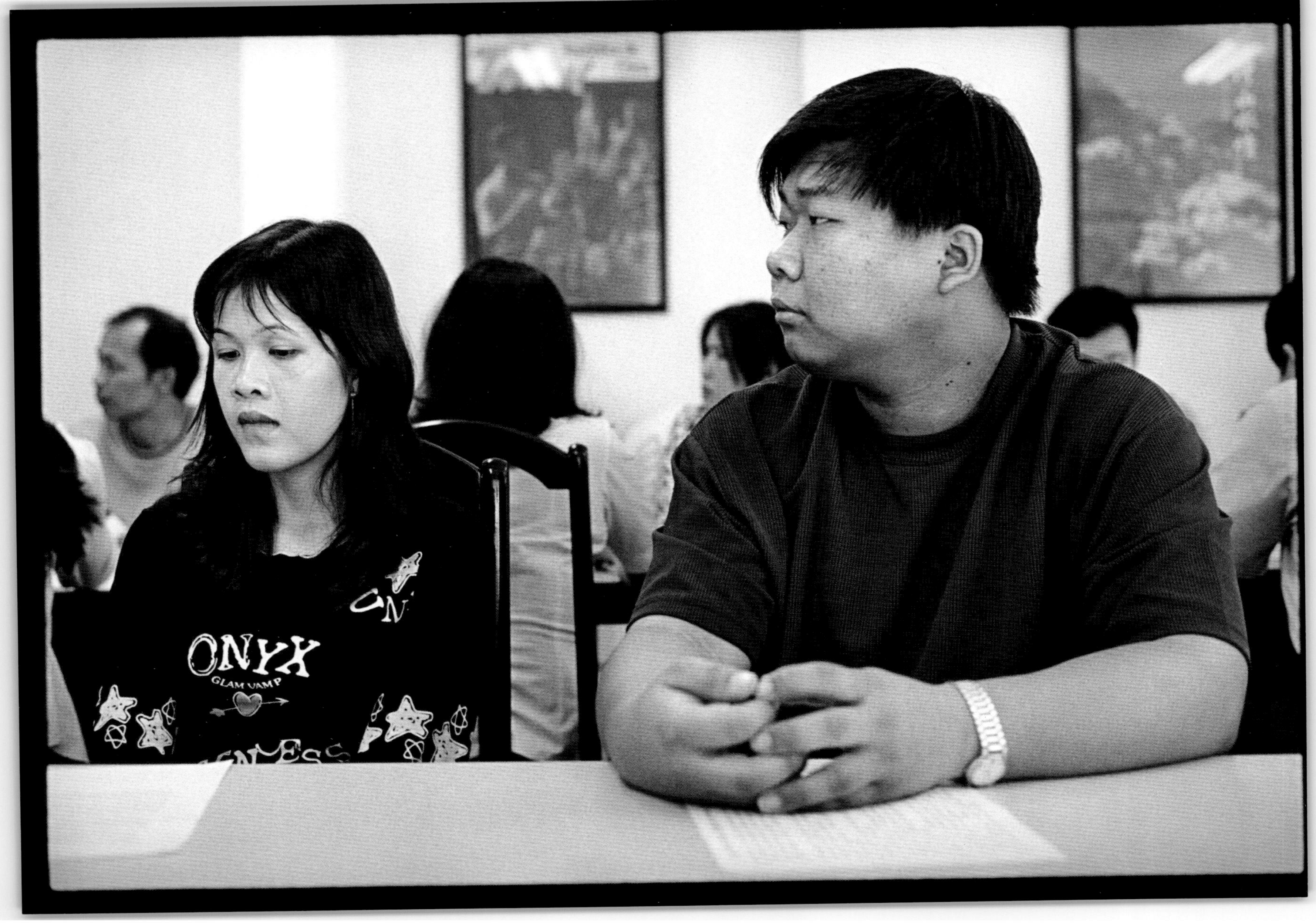
ONYX
GLAM VAMP

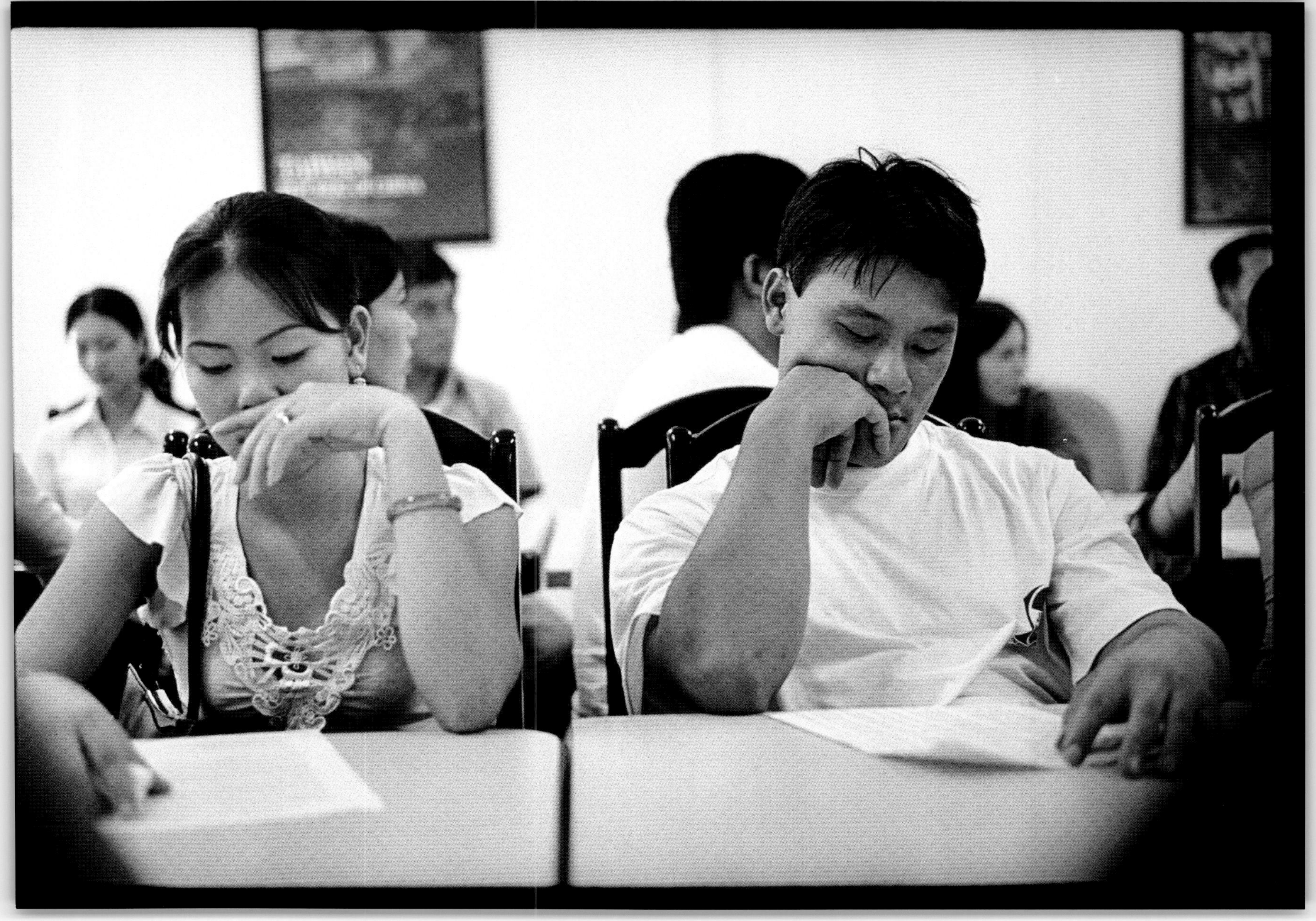

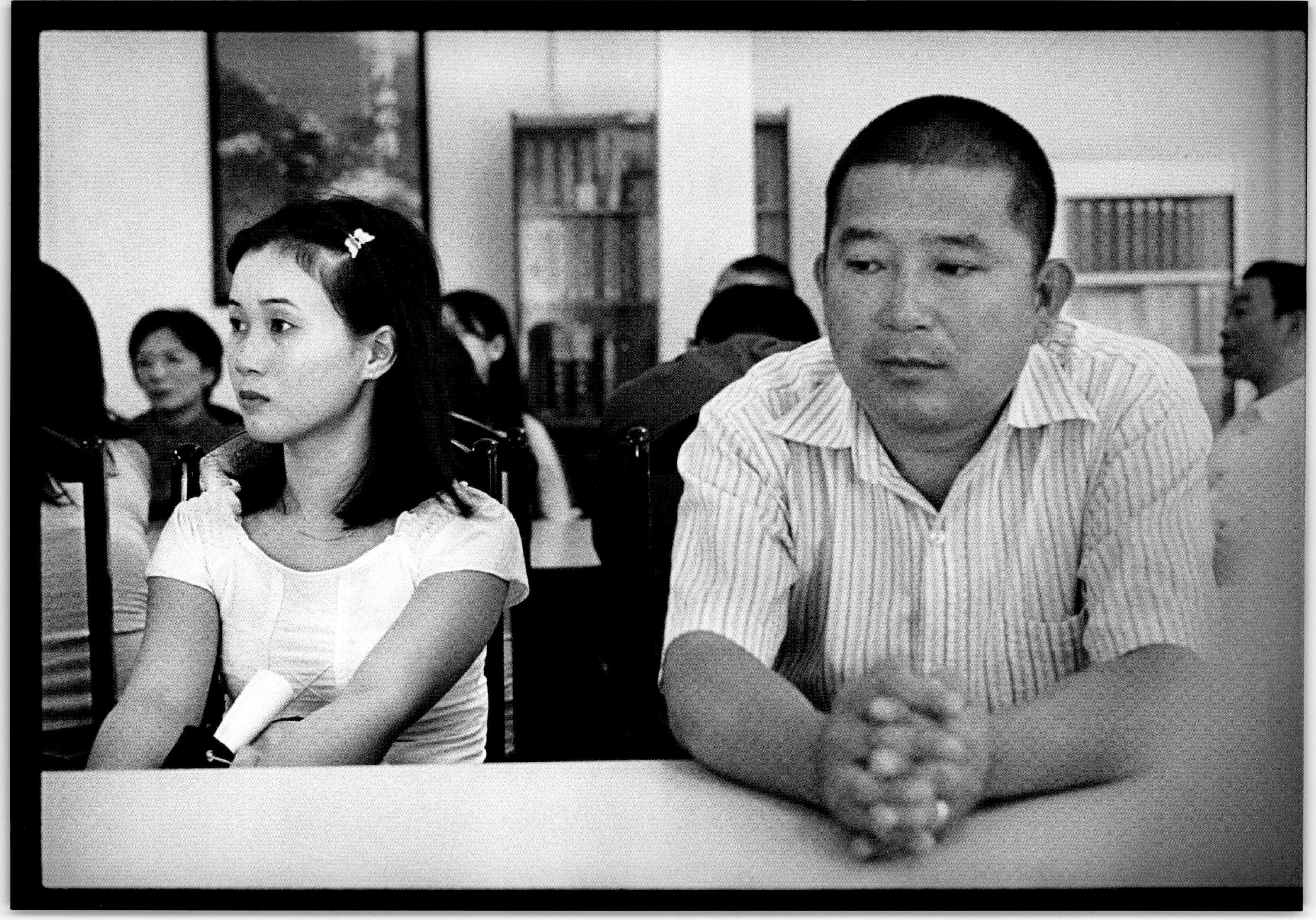

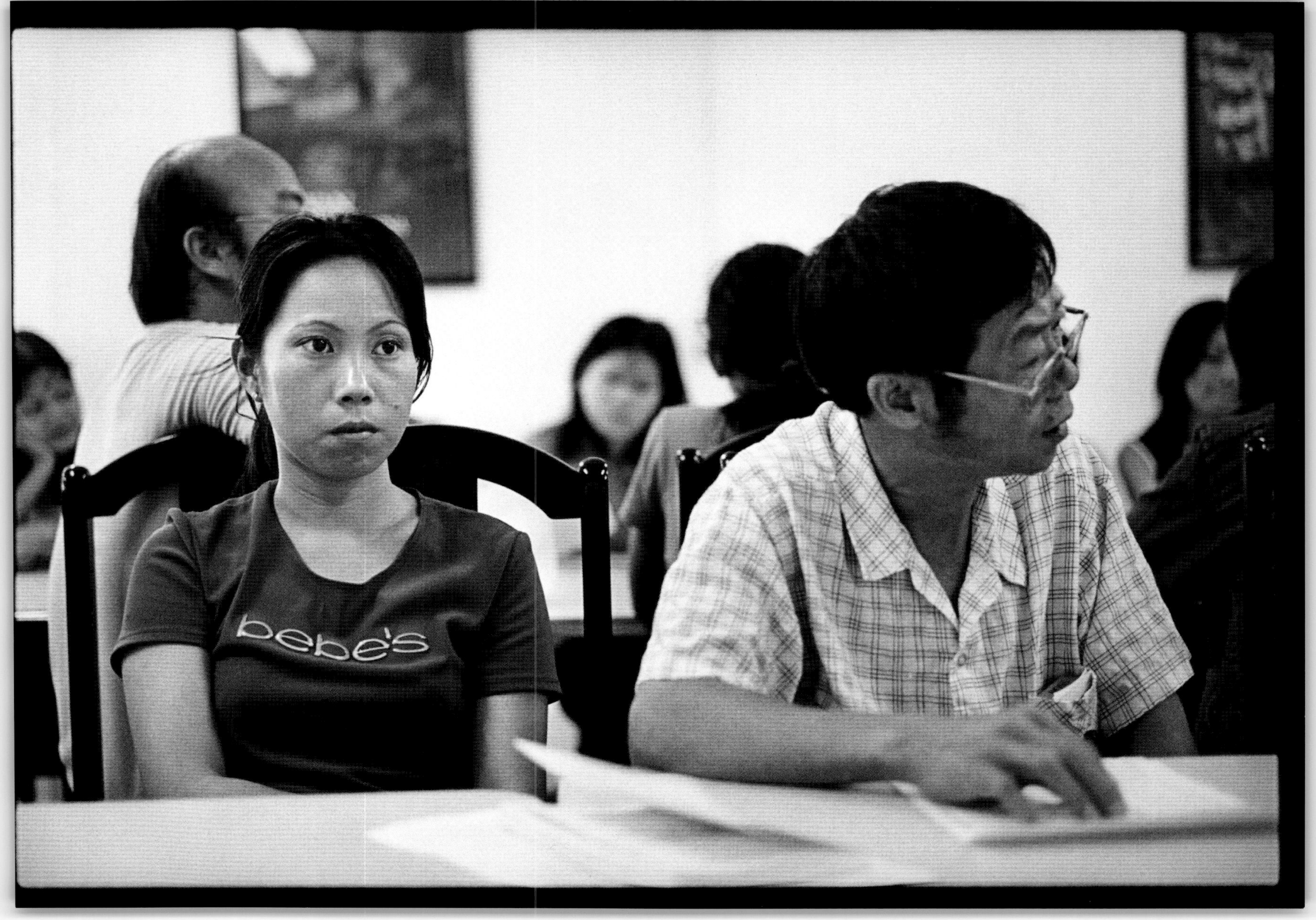
bebe's

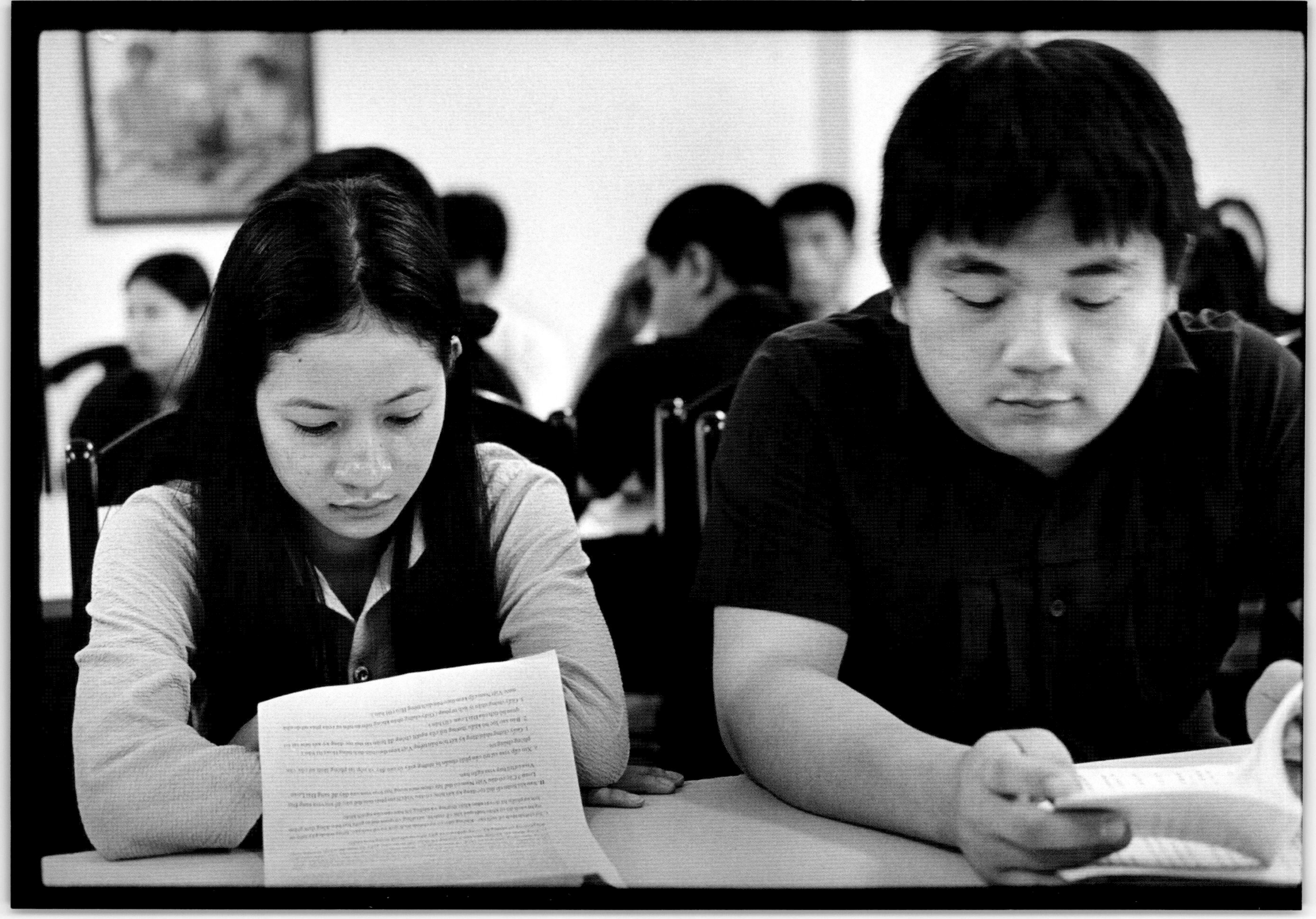

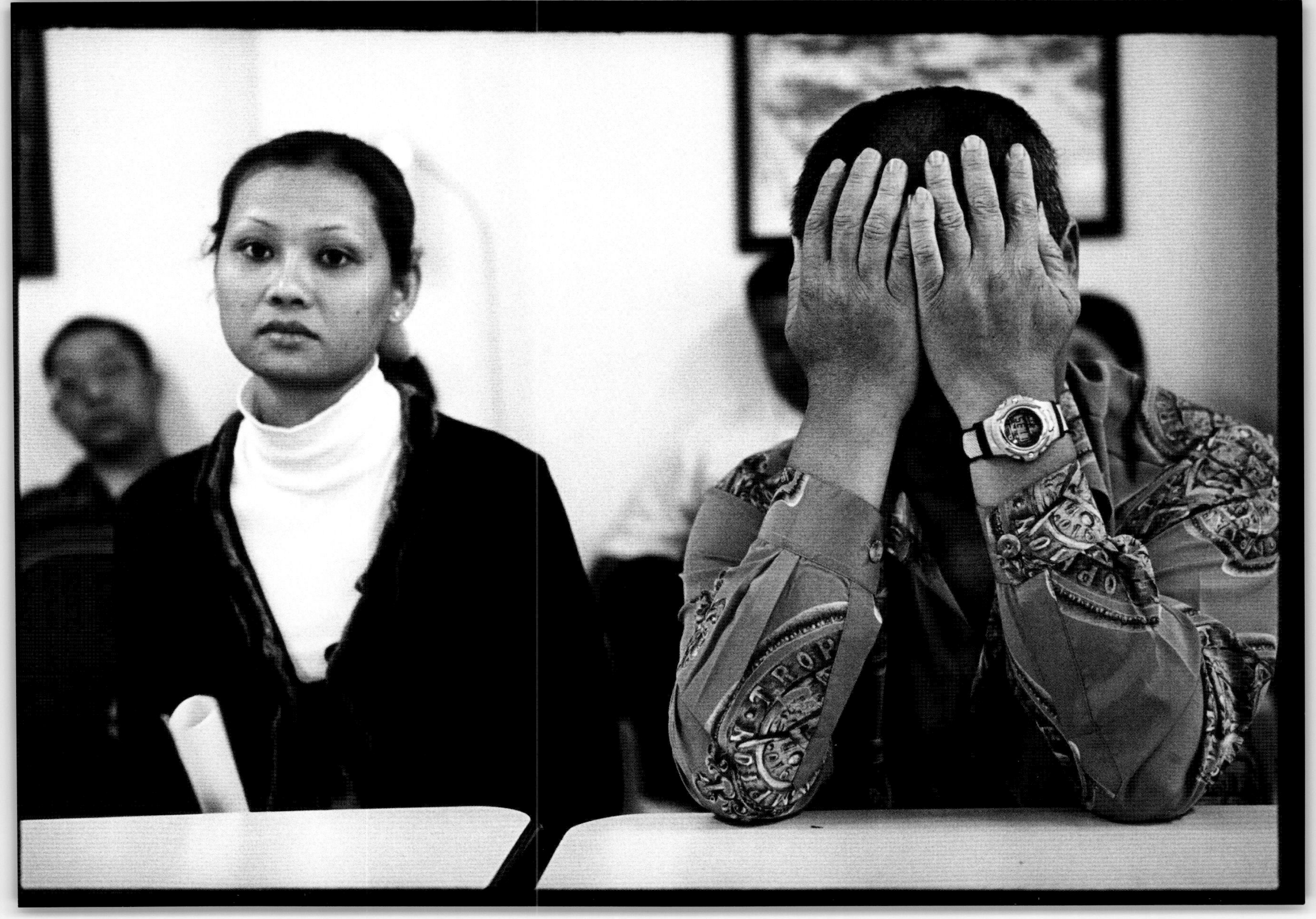

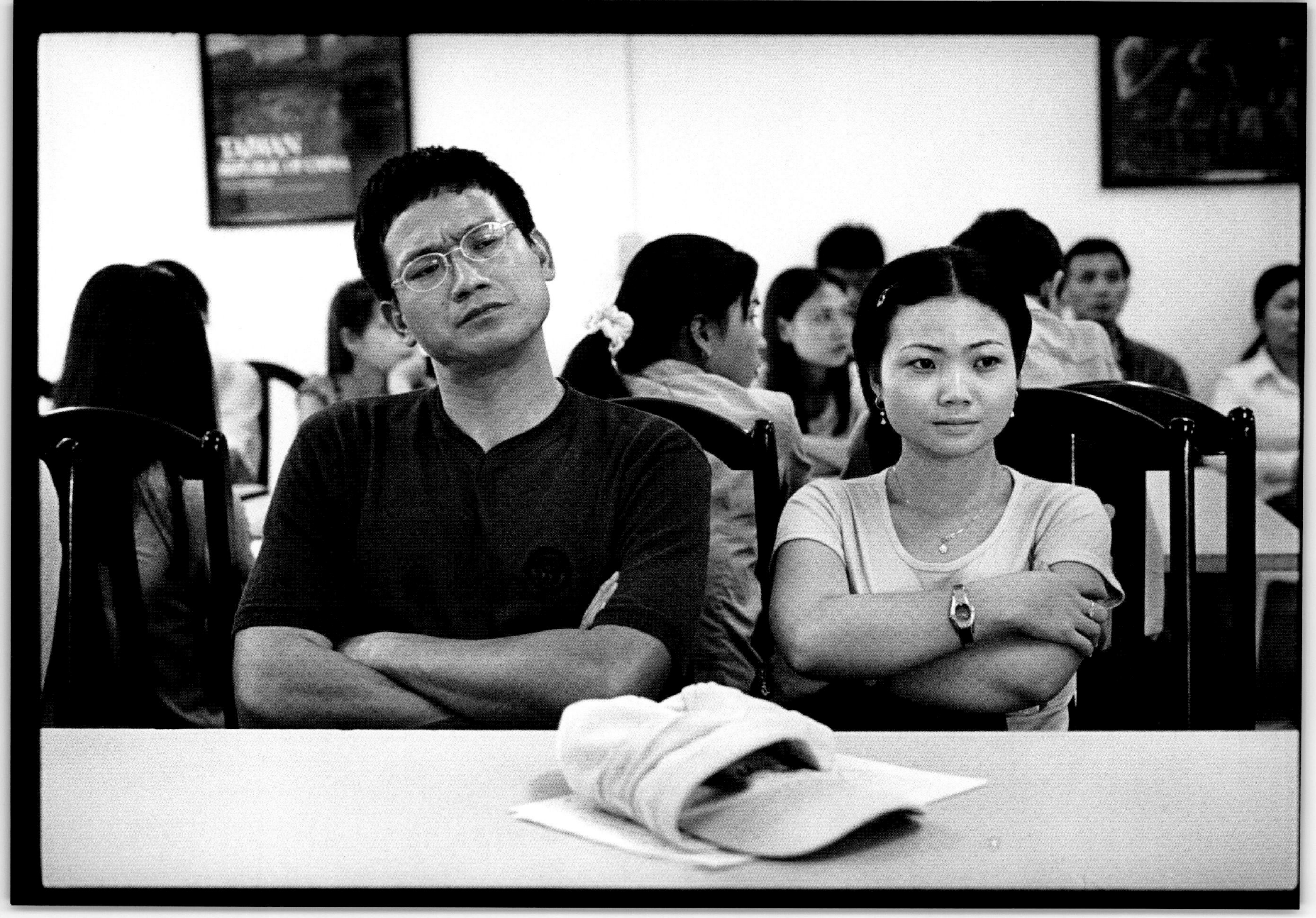

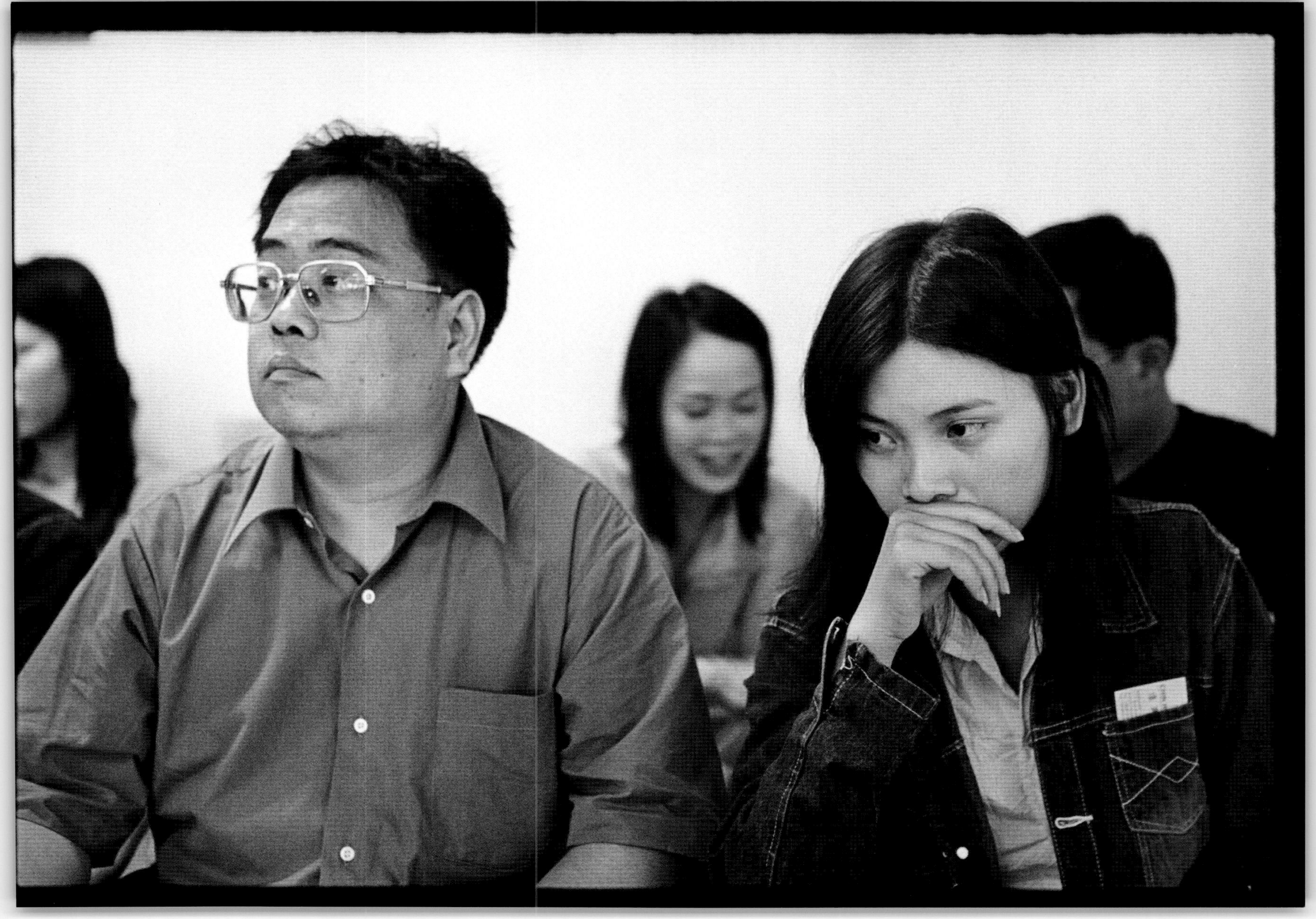

Wedding banquets are held at an amusement park on the outskirts of Ho Chi Minh City; the nuptials are formalized with photo sessions and fake champagne.

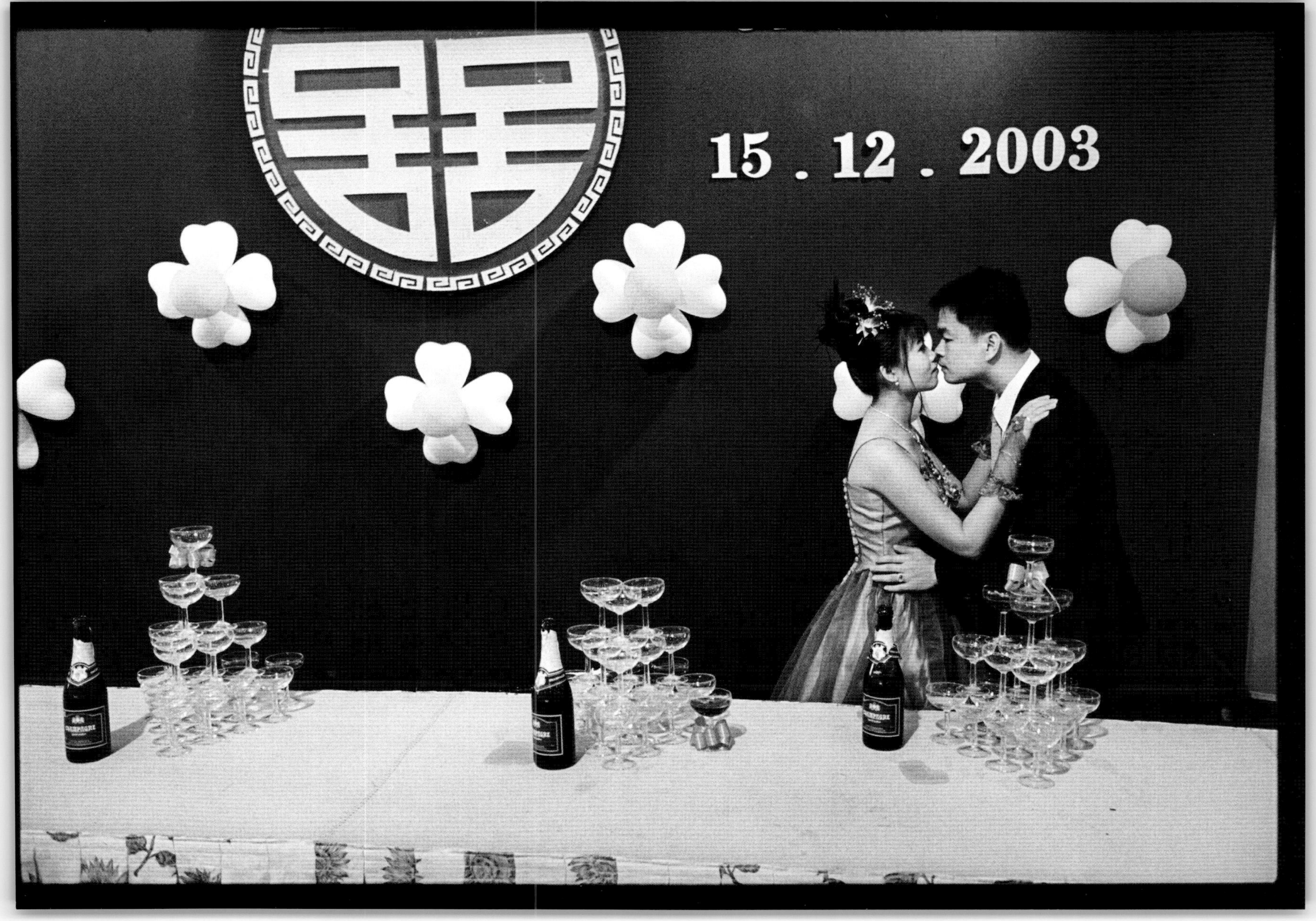
15 . 12 . 2003

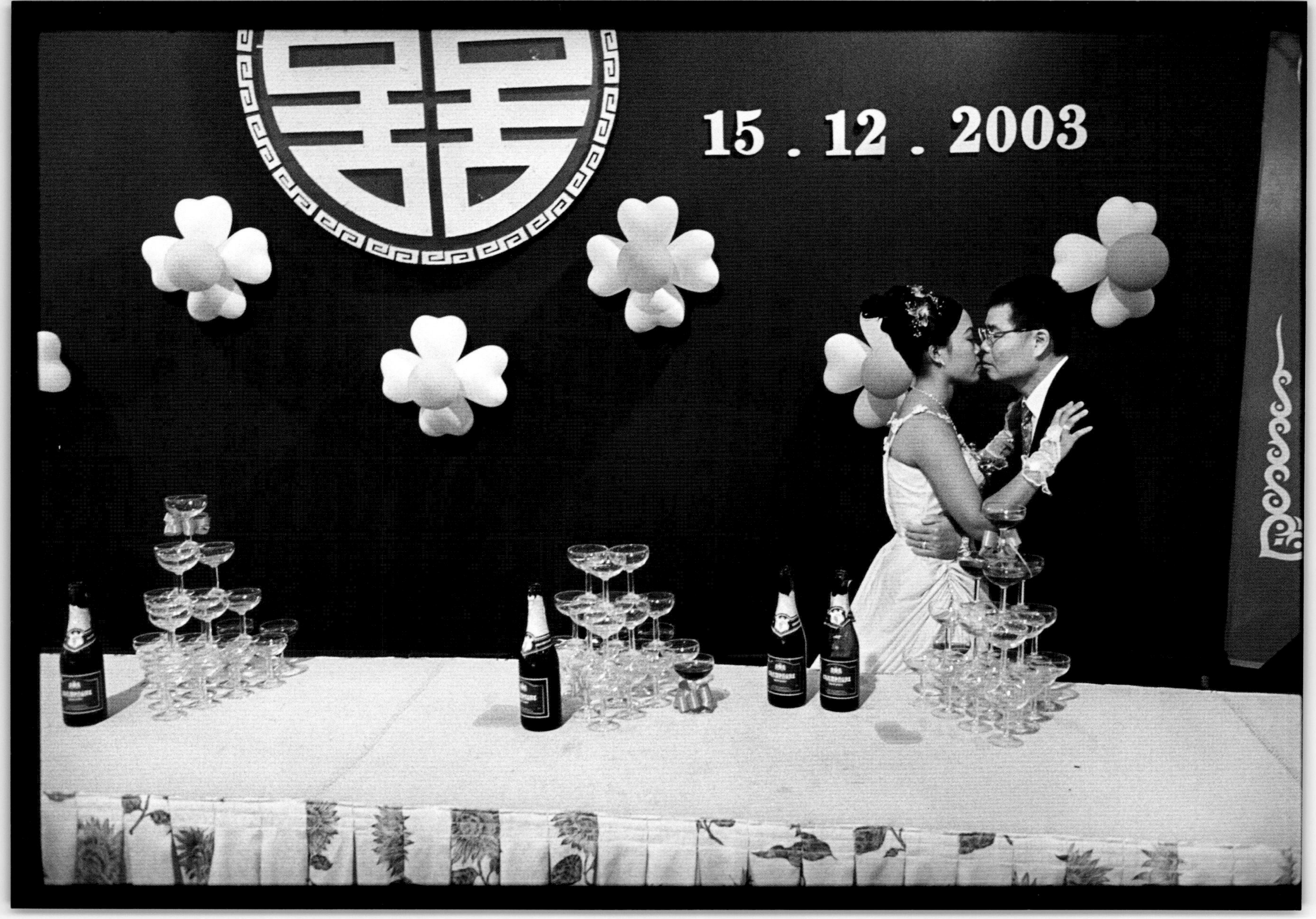
15 . 12 . 2003

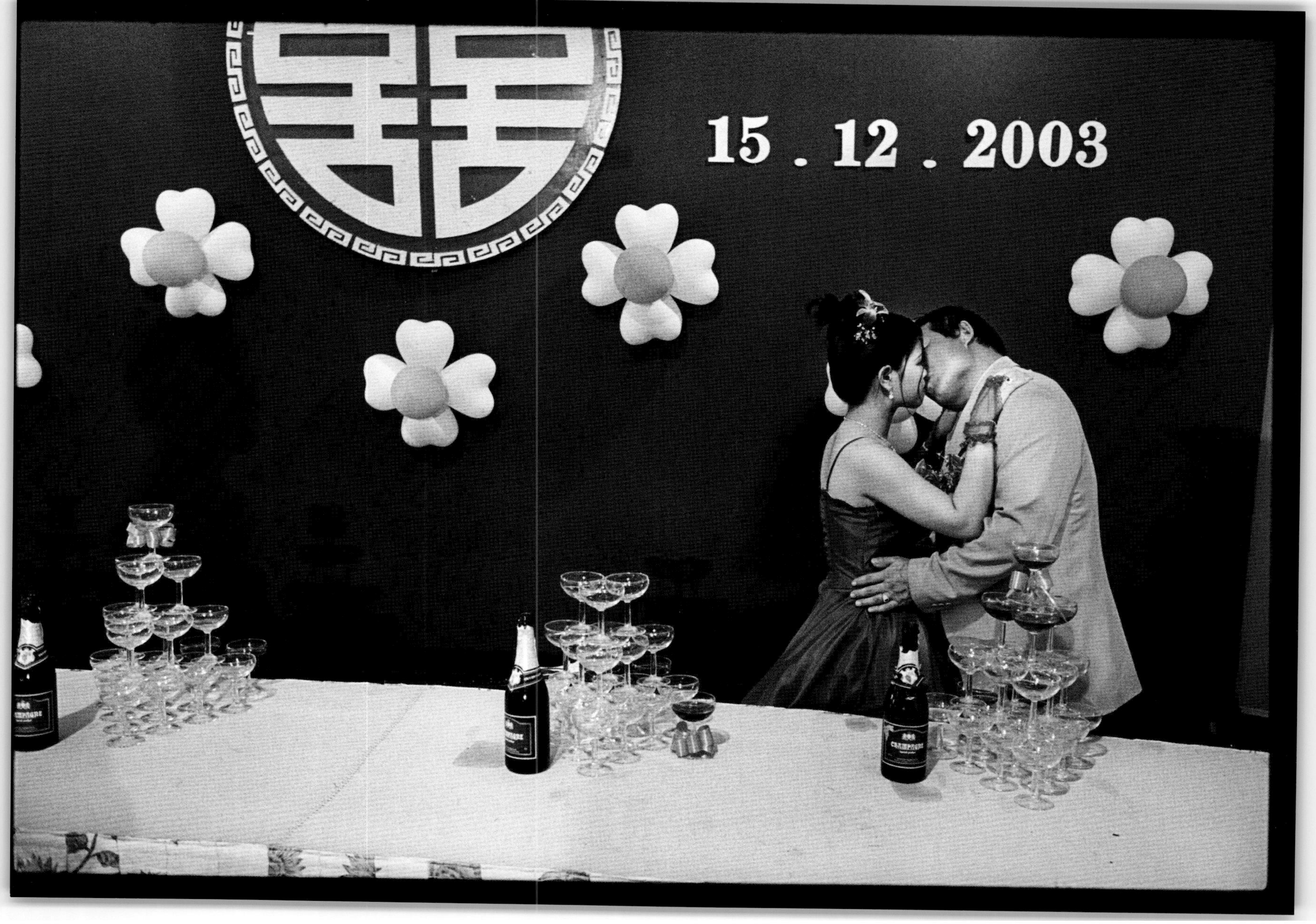
15 . 12 . 2003

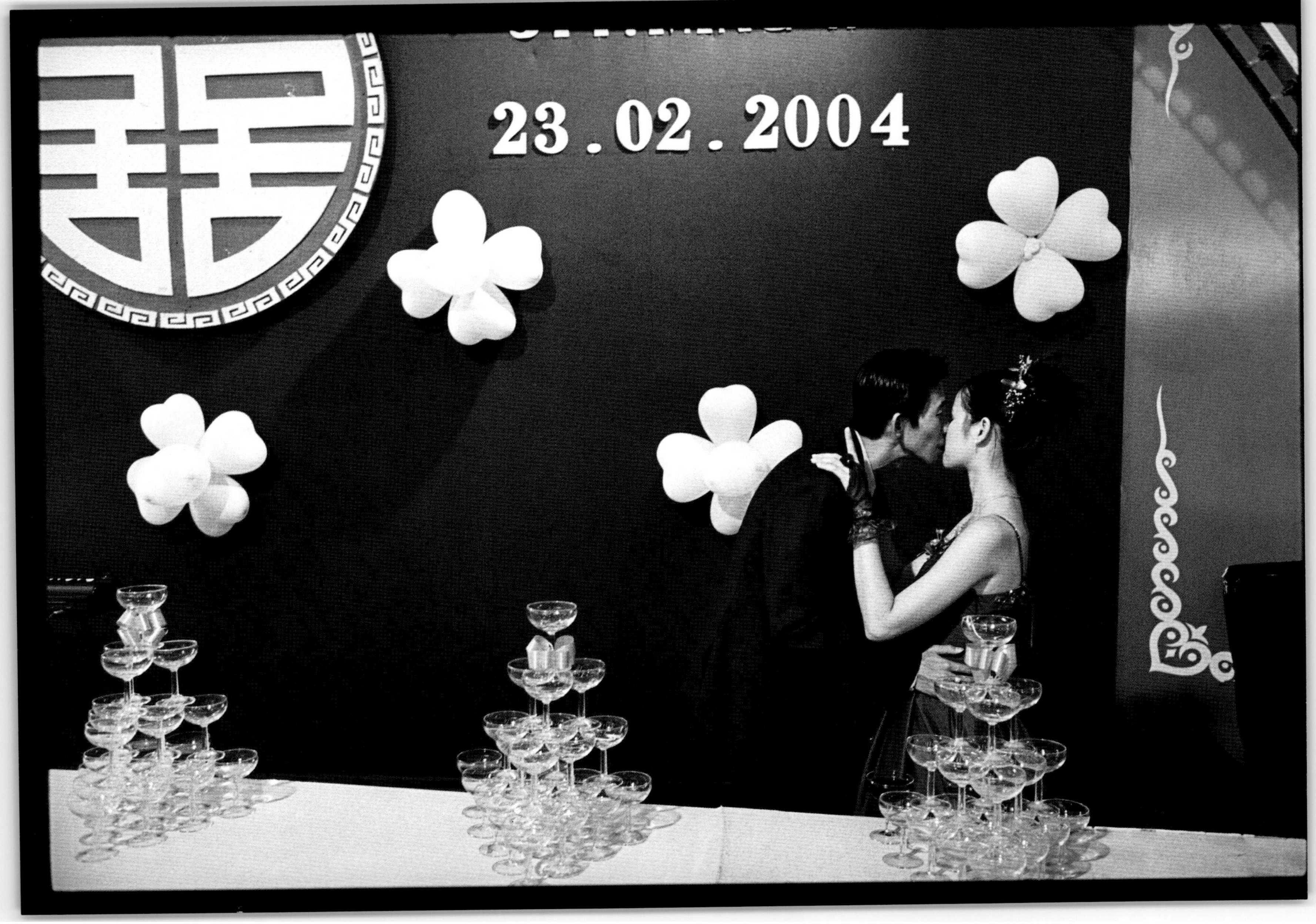
23.02.2004

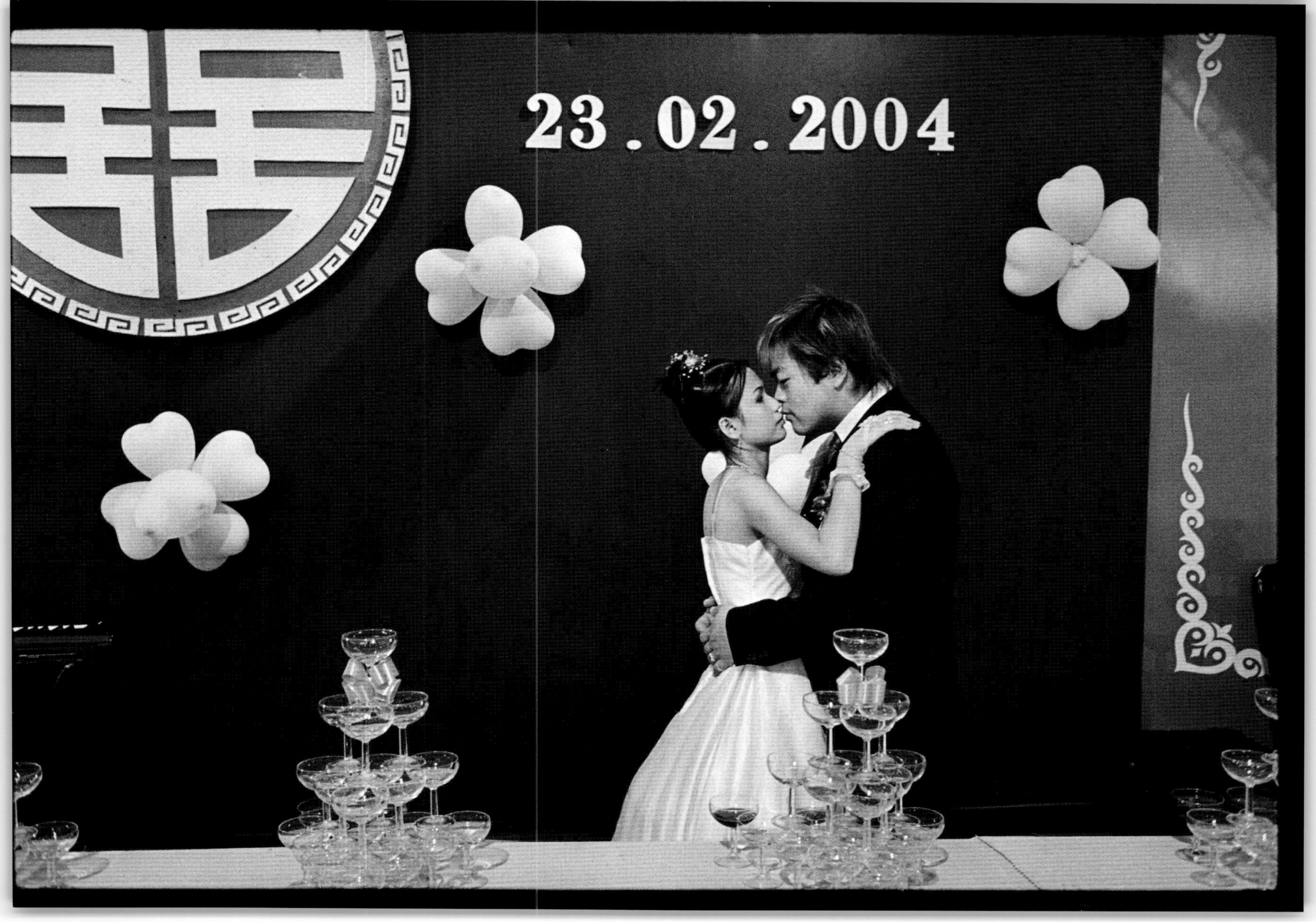
23 . 02 . 2004

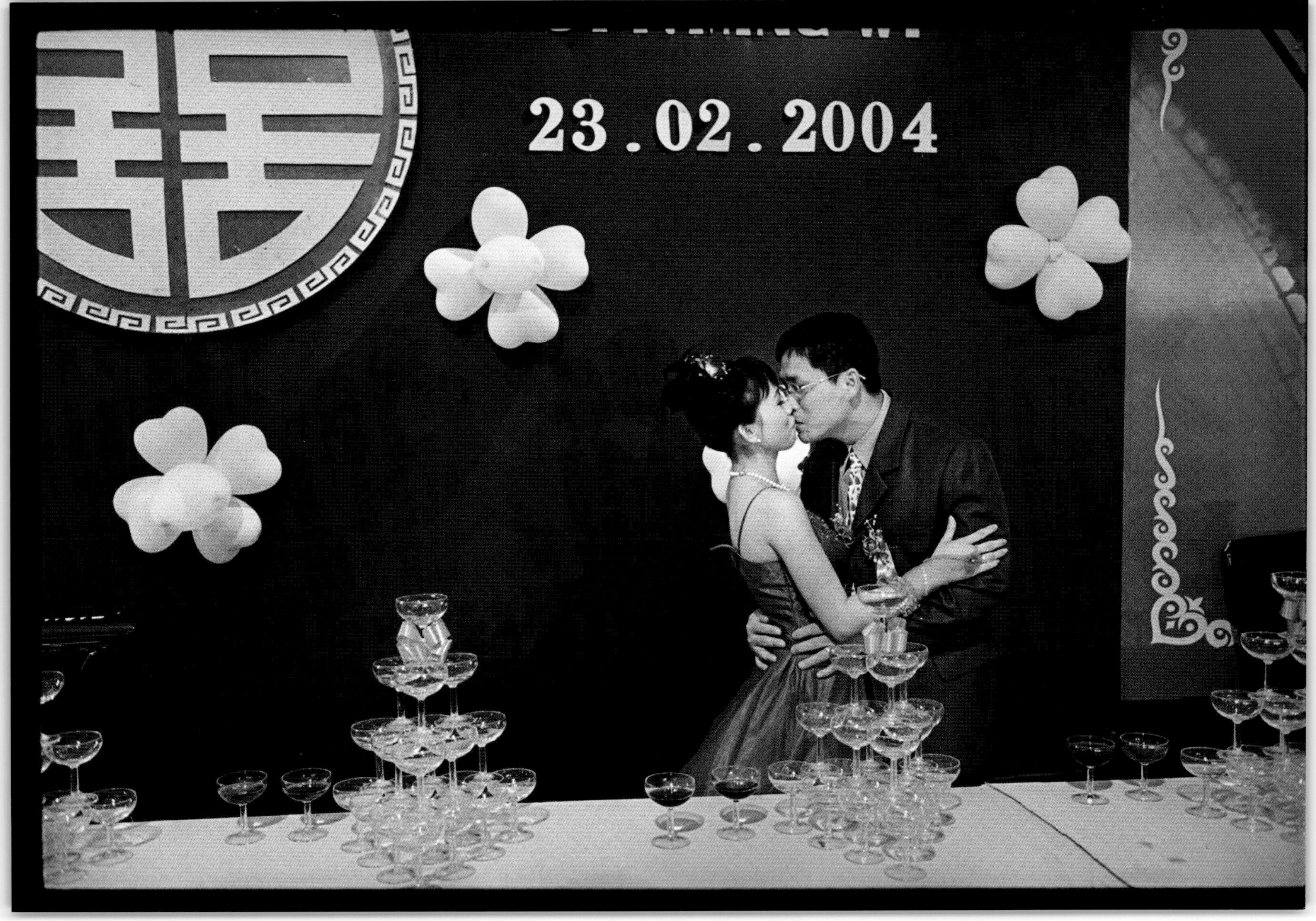
23.02.2004

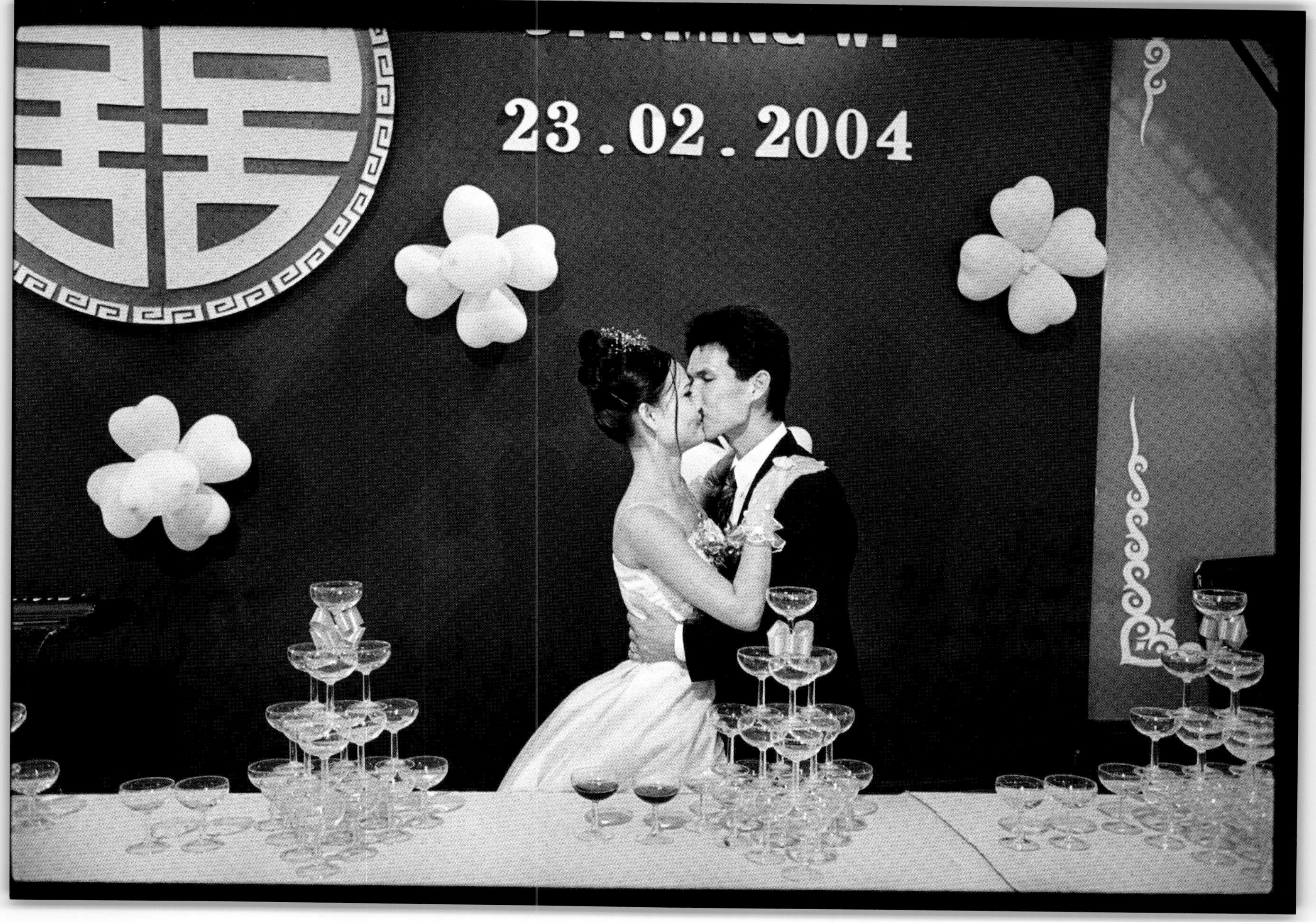
23.02.2004

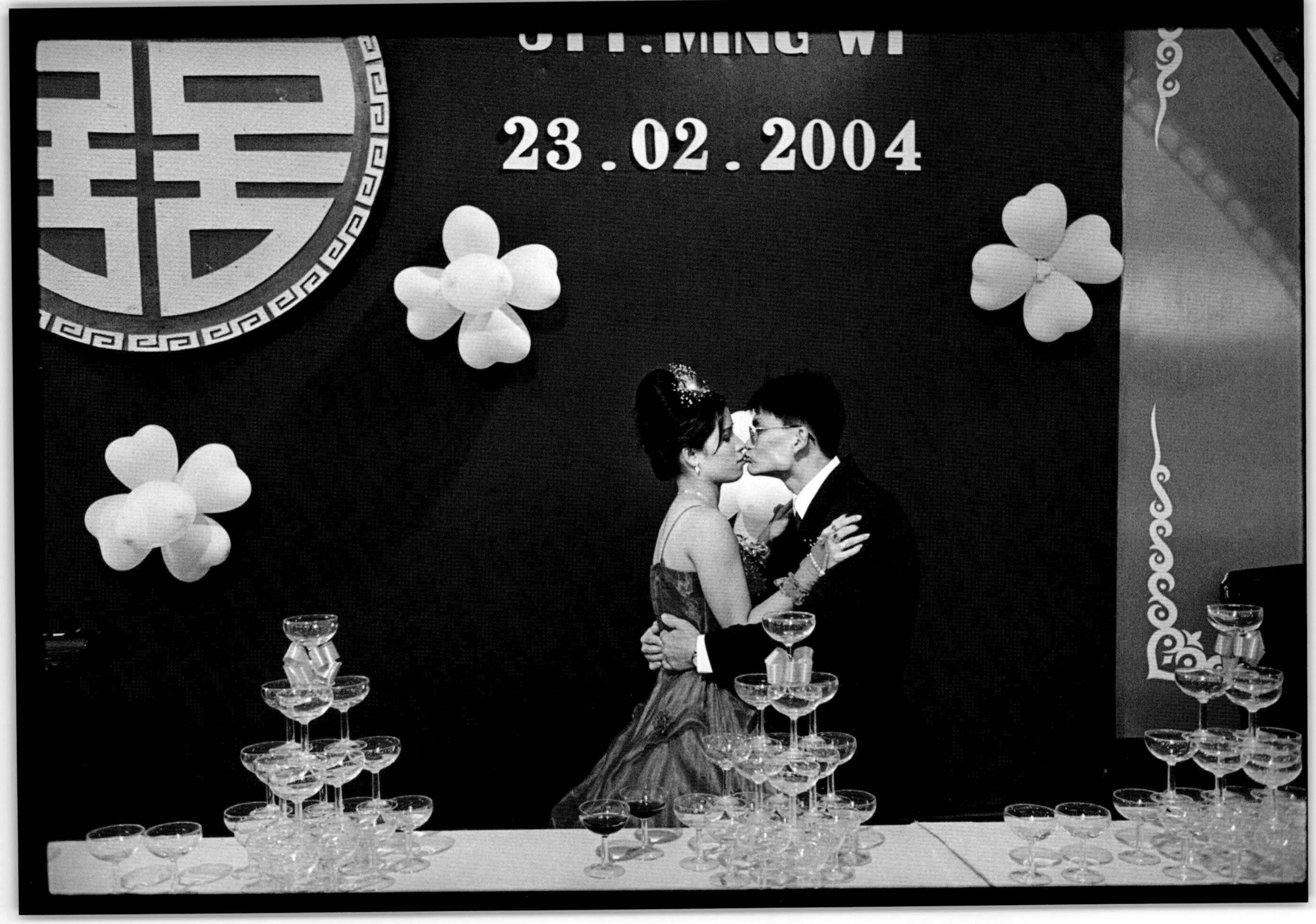
23.02.2004

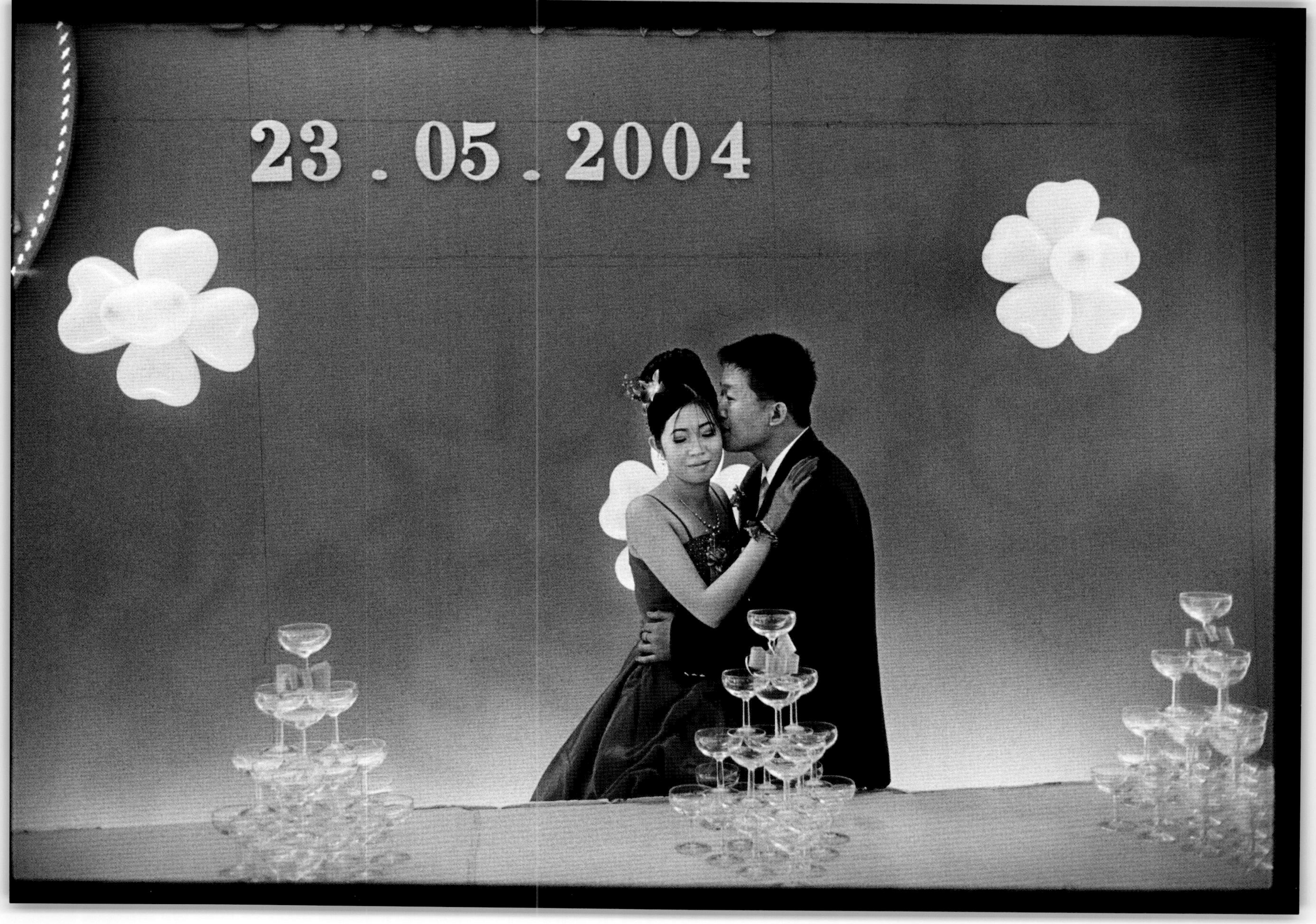
23 . 05 . 2004

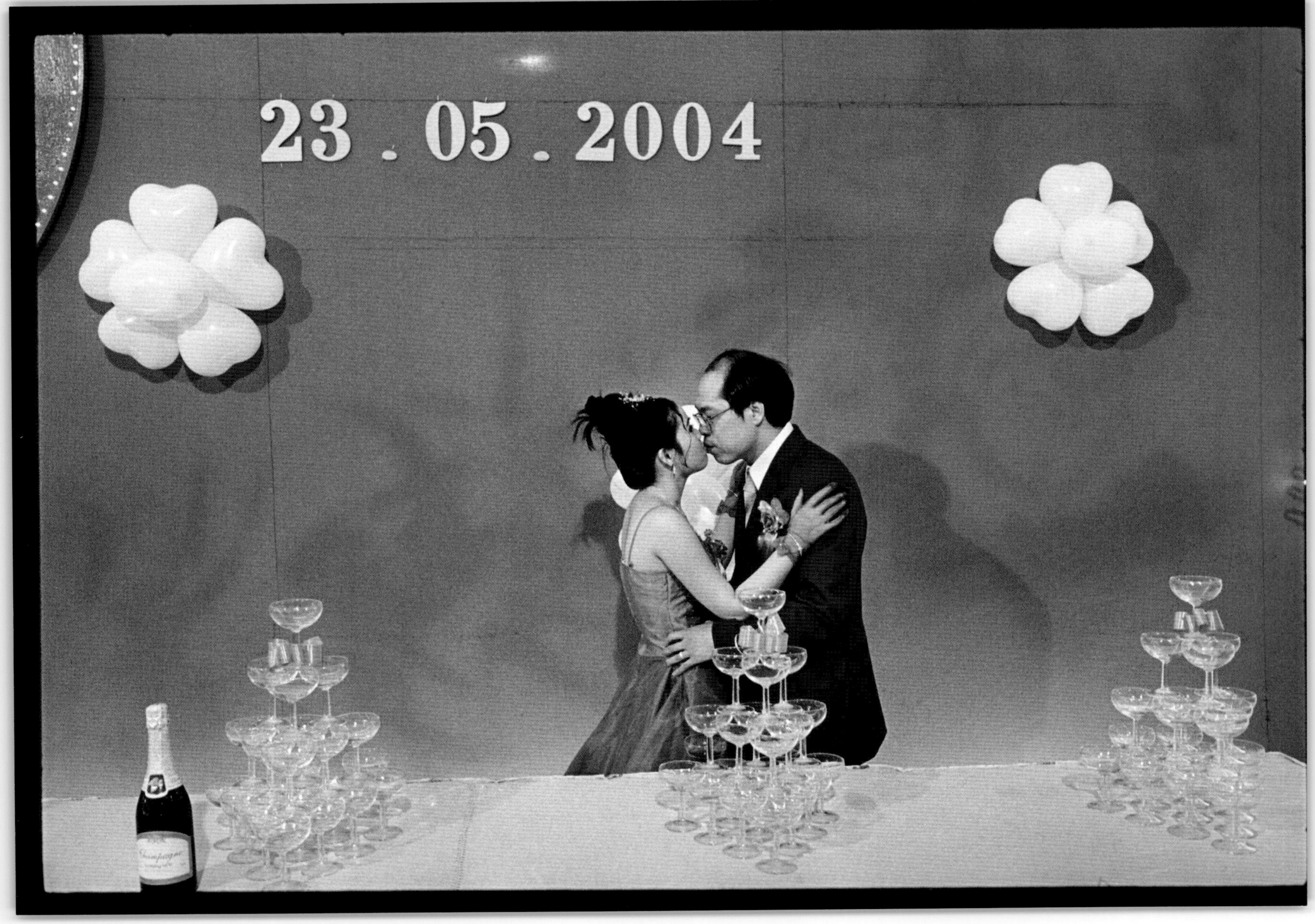
23 . 05 . 2004

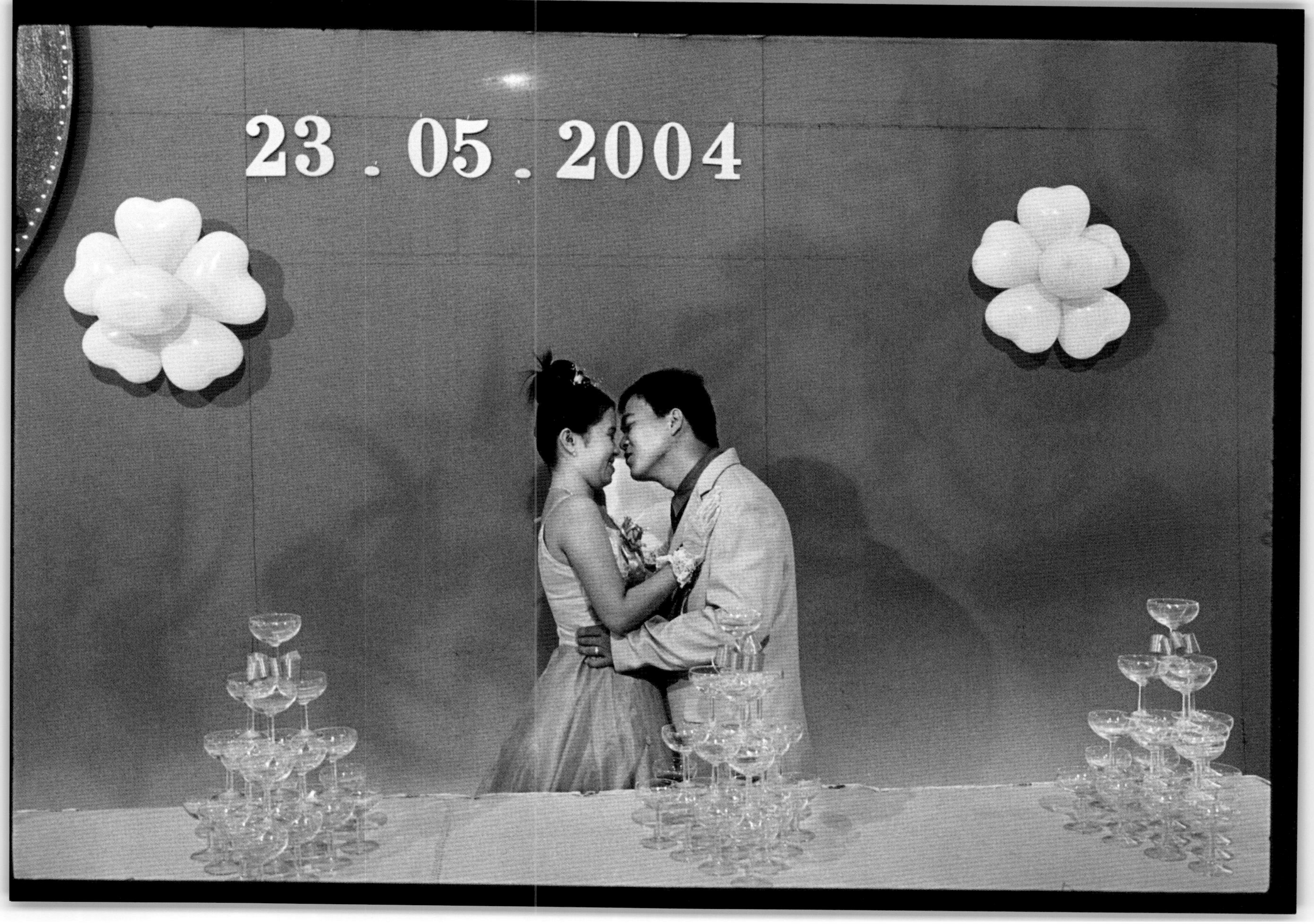
23 . 05 . 2004

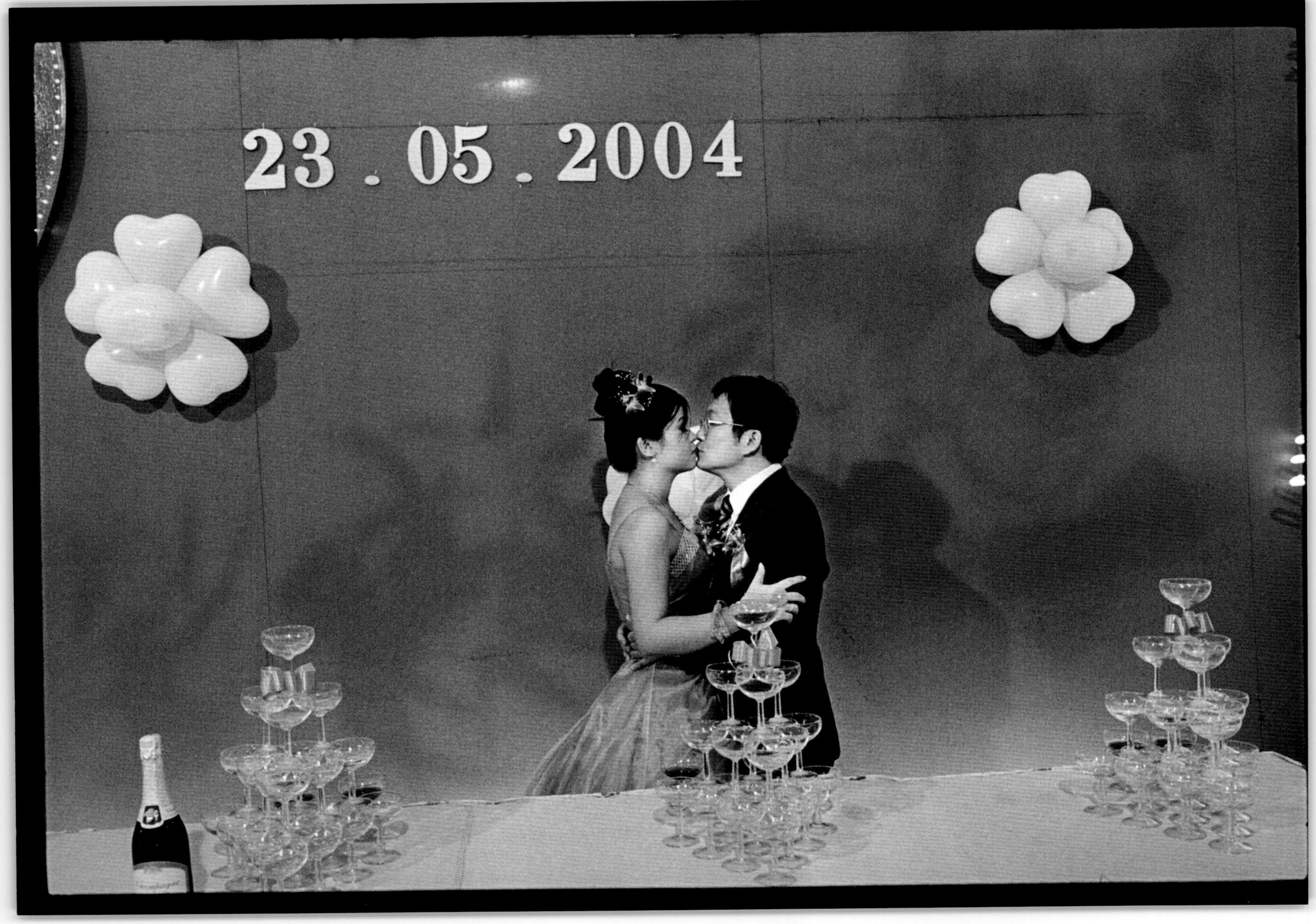
23 . 05 . 2004

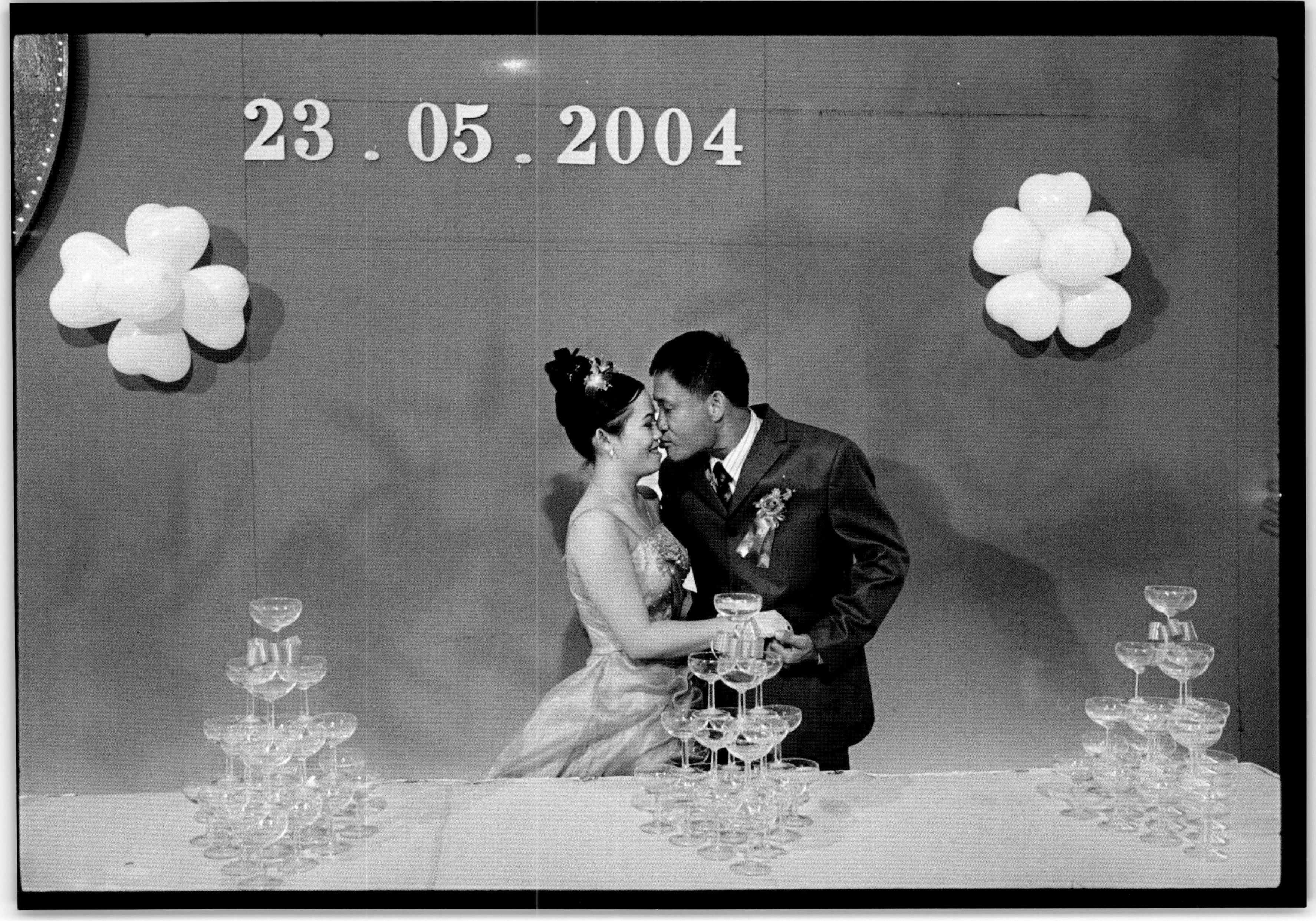
23 . 05 . 2004

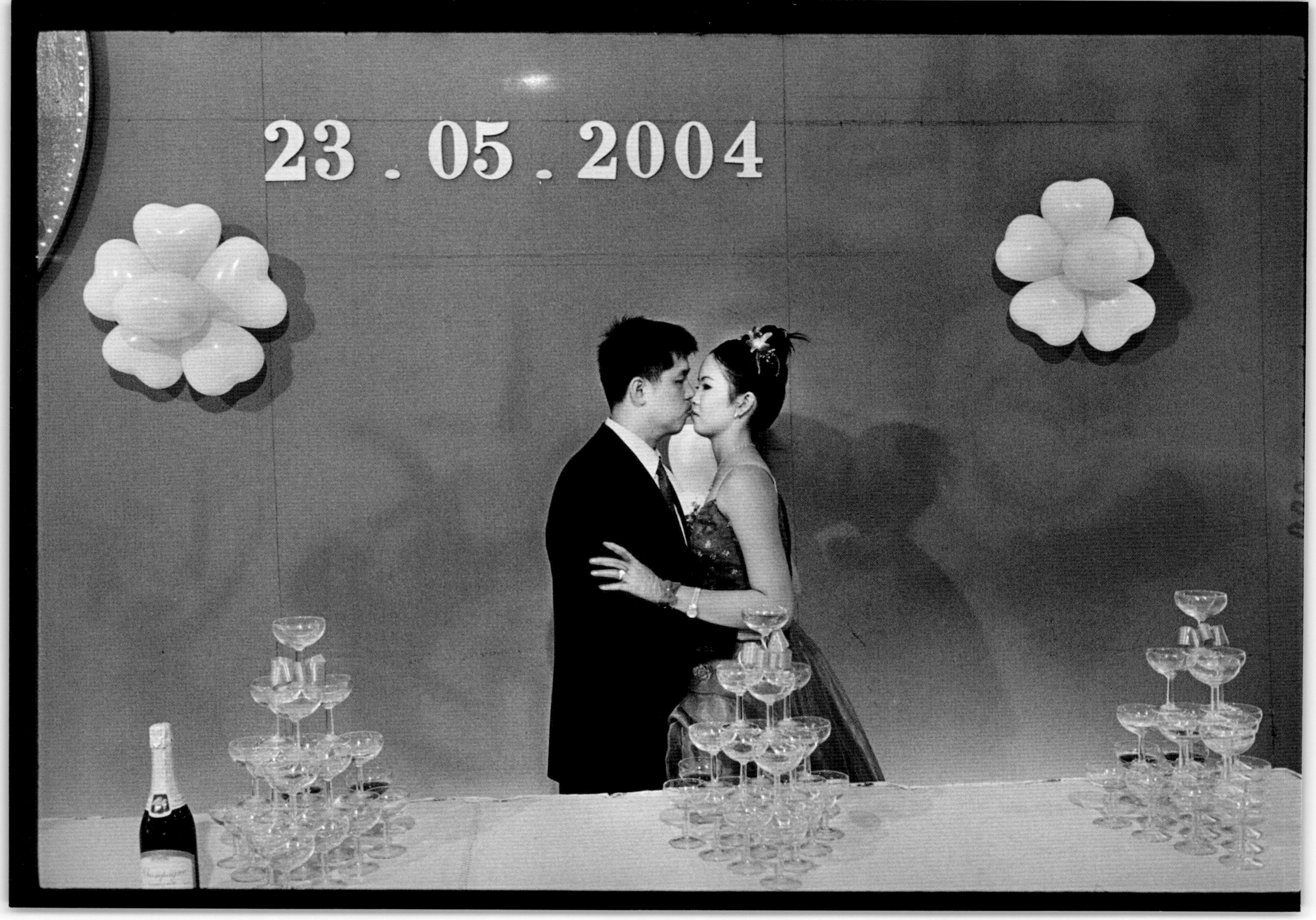
23 . 05 . 2004

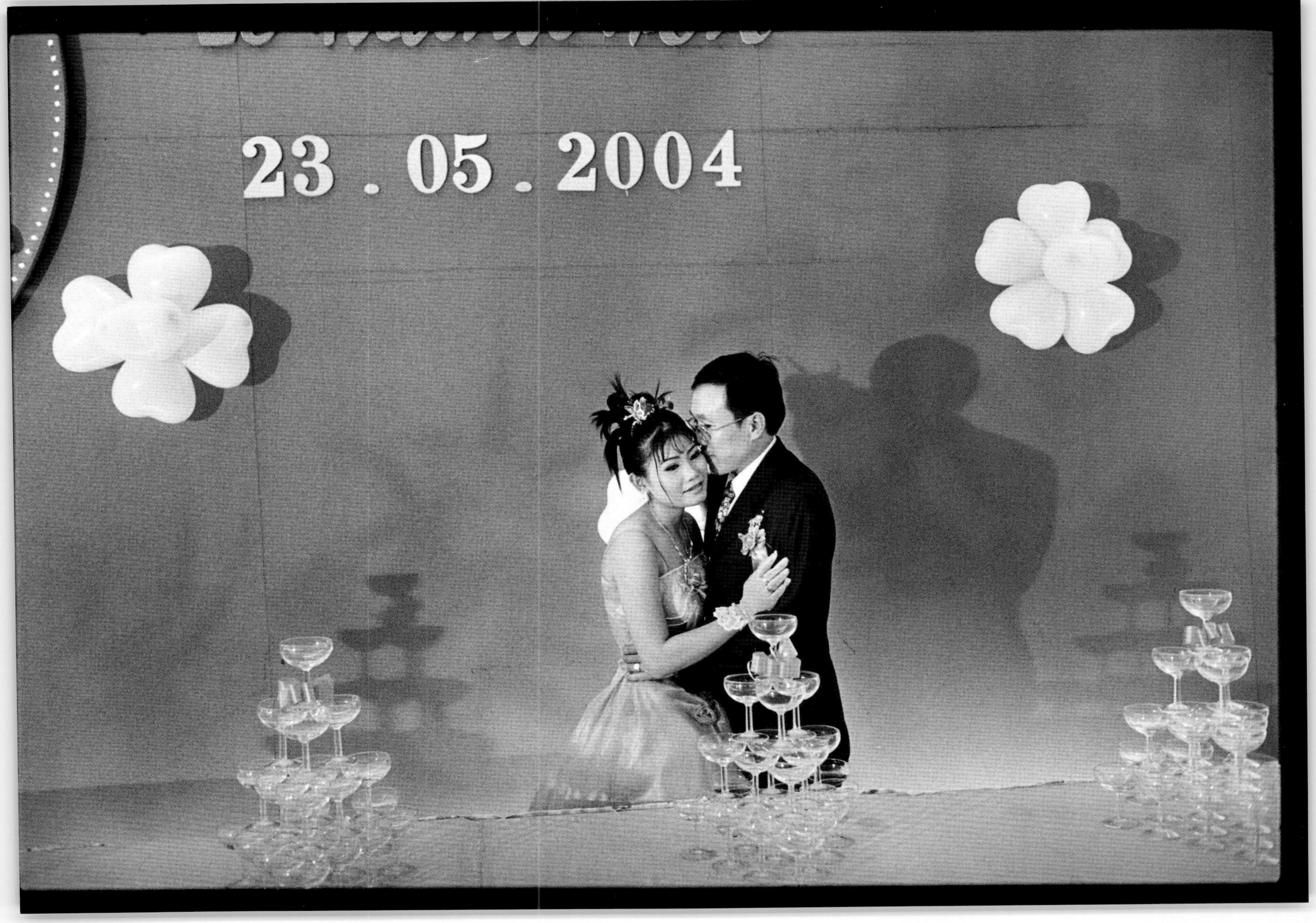
23 . 05 . 2004

the pursuit of happiness by claudia glenn dowling

The man fiddles with a package of Double Happiness cigarettes emblazoned in red with the twinned Chinese character for happiness, the traditional symbol of marital union. He would like to expansively offer the pack to the others sitting around the table, but smoking is not permitted in the Taipei airport café. It is seven o'clock in the morning, and the men are waiting for a plane to Vietnam. A chubby man concentrates on some eggs in front of him. A man with teeth stained reddish brown from chewing betel nuts, a mild narcotic, jokes too much and laughs too loud. A tall, handsome kid stares vacantly as his grandmother knots his necktie for him. She wants him properly attired—after all, he's on his way to meet his future bride. Each of these men is in the market for a wife; by this afternoon, each will have one. Arranged marriages are business transactions burdened with dangerous dreams of love and happiness. Double happiness or double trouble—any marriage can go either way. Yet most are not entered into in just two hours based on a single meeting.

"Hundreds of high-quality beautiful women for men to choose from," scream the billboards and brochures for Lotus 2000. The company is among the most reputable of many that offer matchmaking services to Taiwanese men in search of young, malleable brides. In the past decade, some 80,000 have married women in Vietnam and taken them home to breed and bear, to work and wait on mothers-in-law—tasks that educated, highly paid Taiwanese women are less willing to undertake, particularly for a man in a low income bracket. And the number of foreign brides only continues to climb. "The girls in Taiwan are too picky," says David Wu, chief of the Taiwanese Economic and Cultural Office in Vietnam, which verifies the marriages and issues the women's visas. "They don't want to get married, have children—an eighth of the children in Taiwan are born to foreign brides."

Herded by Lotus guides and chivvied by relatives, the prospective bridegrooms touch down in Vietnam's midday sun to board a bus for downtown Saigon, as Ho Chi Minh City is usually known. At the Lotus headquarters, a cattle call has rounded up the more than a hundred young women who wait in the hallway. Daughters of rice farmers and fisherfolk, they have come to the big city from the River of the Nine Dragons, the Mekong Delta of the south. No man would consider marrying a girl from the north, where use of the defoliant Agent Orange during the Vietnam War has led to birth defects. And each of these girls knows she must bear children—to a man she has not yet met, who does not speak her language, who will take her far from her home and everything she knows.

It is two o'clock in the afternoon. Green coconuts pierced with plastic straws sit on a table in front of the kid and his grandmother. On the wall is a Double Happiness poster. The kid, who works at a car wash, doesn't seem happy. Yawning, he hardly looks at the girls who file in and out of the windowless room in a brutal beauty-pageant parade, at first two-by-two, then in threes, fours, fives. His grandmother, eighty, eyes them sharply. The kid, twenty-seven, is her oldest grandson, and she is determined that he marry.

The girls wear blue jeans and T-shirts and platform shoes. They dip their heads in greeting, then stand against a wall, hands clasped in front of their pubic regions. They are asked their ages—most are between nineteen and twenty-two—heights, weights. No one asks their names.

The air conditioner hums. The kid doodles on a piece of paper.

"*Bu yao*," says the grandmother in Chinese, shaking her head at a group of girls. "Don't want."

The translator is more polite. "*Cam anh*," he says to the rejected girls in Vietnamese. "Thank you."

Down the hall, in another room, the chubby guy has an insouciant air. After all, he has been through this process before: He is back for a second wife. Some months ago, he paid Lotus $8,000 for an all-inclusive marriage package (wife, airfares, hotels, paperwork, wedding banquet, photo album), married a girl in Vietnam and took her home to Taiwan. Within two weeks she had stolen all his cash and absconded. But Lotus guarantees its product: "If a bride's learning ability and personality don't meet the standard, the groom can come back for another selection!" Lotus will provide the chubby guy with another bride for free. And though he won't realize it, this time the brokers have chosen for him—a girl who has been thoroughly vetted. And she's a little chubby, too.

In the next room, the betel-nut chewer, thirty-eight, claims he's not nervous. "Nervous about what?" he asks, laughing. Despite his bravado, an hour into the process he has broken a sweat. "Is that all the girls?" he asks the broker.

"There are more."

"There are too many already," complains Betel Nut. "I need to take a leak. I've had three glasses of water. I feel hot all over."

Three more girls come in.

"No," says Betel Nut.

"Well, the first one's not bad," says the broker. "Fair skin."

"Oh, it's very hard to choose. I'm dizzy. What time is it?"

"It's three o clock—you don't want to look at any more?" asks the broker. "You take a break."

"I'll smoke a cigarette," says the betel nut guy. "I get confused—I must have seen fifty girls.

"You have seen more than a hundred. Three at a time. Thirty runs."

"Is the guy in the other room still picking?"

"Yes."

"I need to pick a good one to come back and help me with work," Betel Nut says. "Water, water. Give me some water."

The kid's grandmother has narrowed down the selects to three. The kid himself has not made a single choice.

"She's too short for him," says the grandmother, who is even shorter than the girl she indicates. The kid's aunt jumps up and drags the kid to stand, sheepishly, next to the three girls. At 6'2", he towers over them. In a wife, tall is preferred, fair is preferred, plump is preferred. Virgin is mandatory. The aunt makes the girls kick off their platform shoes. Like a fortuneteller, she looks at their palms, checking for lines of fidelity and childbearing potential, looking for calluses to prove they can work. She doesn't check their teeth.

"That one's too skinny," the grandmother says, pointing.

"Don't let that be a concern," advises the broker. "When she goes to Taiwan, you can fatten her up."

The kid says nothing. "Don't be shy," says the broker.

A Lotus employee comes in to whisper that a physical examination has revealed that one of the girls is no virgin—she has given birth. The girls are examined for evidence of childbirth or disability. The grooms are examined for mental fitness, questions that can probably be answered even by this kid: "What is 100 minus three?" "Who is your president?" "How many people are in your family?"

The girl who has given birth leaves quickly. The remaining two stand, waiting. "Which one do you like?" the old woman asks her grandson. "You have to make a decision."

The kid giggles.

The broker flips a coin. "Touch their hands and feel the electricity," he suggests. "The one you like, keep holding her hand. The one you don't, let go her hand."

The kid stands between the two girls, holding a hand of each. He drops one hand. The rejected one leaves. The chosen one looks a lot like the grandmother. The new couple poses for photographs in front of the Double Happiness poster.

The chubby guy, who works for a trucking company, has been tactfully influenced: He picks the chubby girl. The girl seems relieved—she has likely been on the market for a while. A girl is allowed to reject suitors, but if she turns down too many, brokers may start pressuring her to accept an offer—after all, they foot the bill for the girls' room and board in the city. While at Lotus's center, the chubby girl studied Chinese language. She also studied Chinese cooking—the chubby guy, like most of the men, does not like Vietnamese food.

Betel Nut is getting closer to his final selection. The broker consults his notes. "You've picked twelve," he tells Betel Nut. "Now eliminate the ones you don't like."

The finalists file in four at a time.

"OK, keep number three," says Betel Nut. "Oh, now it's difficult to make a choice. The red shirt one and the black shirt one. Let me go to the bathroom."

The finalists' heads swivel to watch him as he leaves the room. While he's gone, the broker fills the girls in. Betel Nut drives a truck for a paint factory. He lives in an industrial suburb of Taipei with his retired father. His mother is dead. One girl asks what his salary is. The broker says he makes about $1,060 a month, but shakes his head at the question. The girl is too grasping. The average income in Vietnam's countryside is about $30 U.S. a month, compared to Taiwan's minimum wage of $475. These matches are economic shotgun weddings—a girl marries a stranger to improve her standard of living and send money home to her family. Still, if they seem too eager for the money, the Taiwanese are put off.

Betel Nut comes back. One of the girls nervously tugs down her red shirt. She smiles. She has a crooked front tooth. She tells the translator to tell Betel Nut: "If I'm married to you, and you're at work all day, I'll be alone, and I'll be scared."

"If I marry her, I'll be working as a driver, and I'll bring her along," says Betel Nut. "I will be worried, too, if I leave her alone."

The girl nods.

"She has a smiling face," muses Betel Nut.

"The first one has a smiling face," agrees the

Lotus boss, who shows up to close the deal. "You may not be a driver forever. Maybe you will open a shop, and she will be an asset with her smiling face."

The brokers question the girls about their families. Each comes from a farm, has an elementary school education. Each hopes to send money home regularly. The one with the smiling face admits that her parents owe their neighbors $300. Lotus will settle $500 on the family if she marries Betel Nut.

The two girls chew their lips. They can't help wanting to "win," to be picked. At the same time, they worry as they face a husband in the flesh, no longer an abstract dream of financial security but a man to wed and bed.

Betel Nut picks the girl with the smiling face, almost two decades younger than he. Will she still smile as she makes the requisite sacrifices to filial obedience? She takes his hand. The other girl leaves the room to wait for the next viewing.

Outside in the hall, Betel Nut runs into the chubby guy with the chubby girl. They, too, are holding hands. "Oh," says Betel Nut, "She was up for grabs, too?"

In two hours, the future of six people has been arranged. By 4:30 the newly betrothed are in a van, headed to the hotel. There is no conversation—the couples don't share a language.

This is the first day of a marriage. What of all the other days and years? Even newlyweds of the same culture often feel like they've married a foreigner. These couples don't share a language. Each partner can become a mannequin to clothe with expectations, a construct of the secret heart of desire. "Men and women are gambling on their happiness," concurs Wu, the Taiwanese official who must oversee the paperwork for nearly 10,000 such couples every year. "They cannot communicate. They don't know the family background, emotional background—nothing! In one day men see fifty girls and pick someone to spend their lives with."

Many such weddings are held three days after the selection—as one broker says, "All the sows are in the sty. It's time to pay!" Lotus, however, requires engaged couples to wait two months. "Some brokers claim that right after matchmaking you can get married and have sex," warns the Lotus brochure. "But what if she has STDs or AIDs? Wait until the second trip to go to bed with your wife. Meanwhile, she will have blood tests to protect you and the next generation."

There are hustlers on both sides. “Men are not told all the risks—perhaps she has a crime record or is married,” says Wu. “The girls also do not know anything. They believe they are like Cinderella and Prince Charming. In Taiwan they have to take care of the children, husband, his parents.” Mothers-in-law, he says, can be a particular problem—it is no accident that the Chinese character for “trouble” shows two women under one roof. “One girl committed suicide, because she did not know she would have such a heavy burden,’ says Wu.

Culture shock compounds the difficulties. “A foreign bride can be a time bomb,” asserts one broker. “And *every* Taiwanese man who comes to look for a bride has a mental problem—the fact that he can choose from among sixty girls gives him an illusion of power.” When a girl finds herself in a strange country, married to a stranger, that power is no illusion. She doesn’t speak the language and is cut off from her support systems. She can’t claim Taiwanese citizenship for three years and may never learn to read and write Chinese. If her husband overworks her or beats her or shops her to his friends as a sex slave, she knows no recourse. The Taiwanese government does man a hot-line number answered in Vietnamese, but many abused brides are afraid of authorities. As many as 10 percent of Taiwanese-Vietnamese couples will divorce, some because of spousal abuse. Compared with a 25 percent divorce rate in Taiwan, however, statistically, that looks like happiness. Statistics don’t measure homesickness, disillusionment, resignation, despair.

The men who met in the airport just three days ago are on their way home. Each has selected a wife, celebrated at an engagement banquet and put the necessary paperwork in process. The kid’s grandmother removes his engagement ring from his finger every night and gives it back to him in the morning; perhaps afraid he’ll lose it. His fiancée is very attentive to grandmother. The chubby guy is ready to give No. 2 a try—as soon as he gets divorced from No. 1, whereabouts unknown. Betel Nut’s fiancée of the smiling face has informed him, despite the language barrier, that she won’t kiss him if he doesn’t stop chewing betel nuts. He is chewing gum. On their next trip to Vietnam, the three men will be married. Double Happiness cigarette, anyone? As they bus to the airport, a broker breaks open a case of bottled water with a label reading “Joy.” There is plenty of bottled joy left in the case for the next group of men.

SEVEN COUPLES GO THROUGH THE PROCESS OF SELECTION, AGREEMENT, AND CELEBRATION.

On August 24, 2003, Nguy Hoai Ai (far right) was selected by Chen Yu-tsen; they were wed four days later.

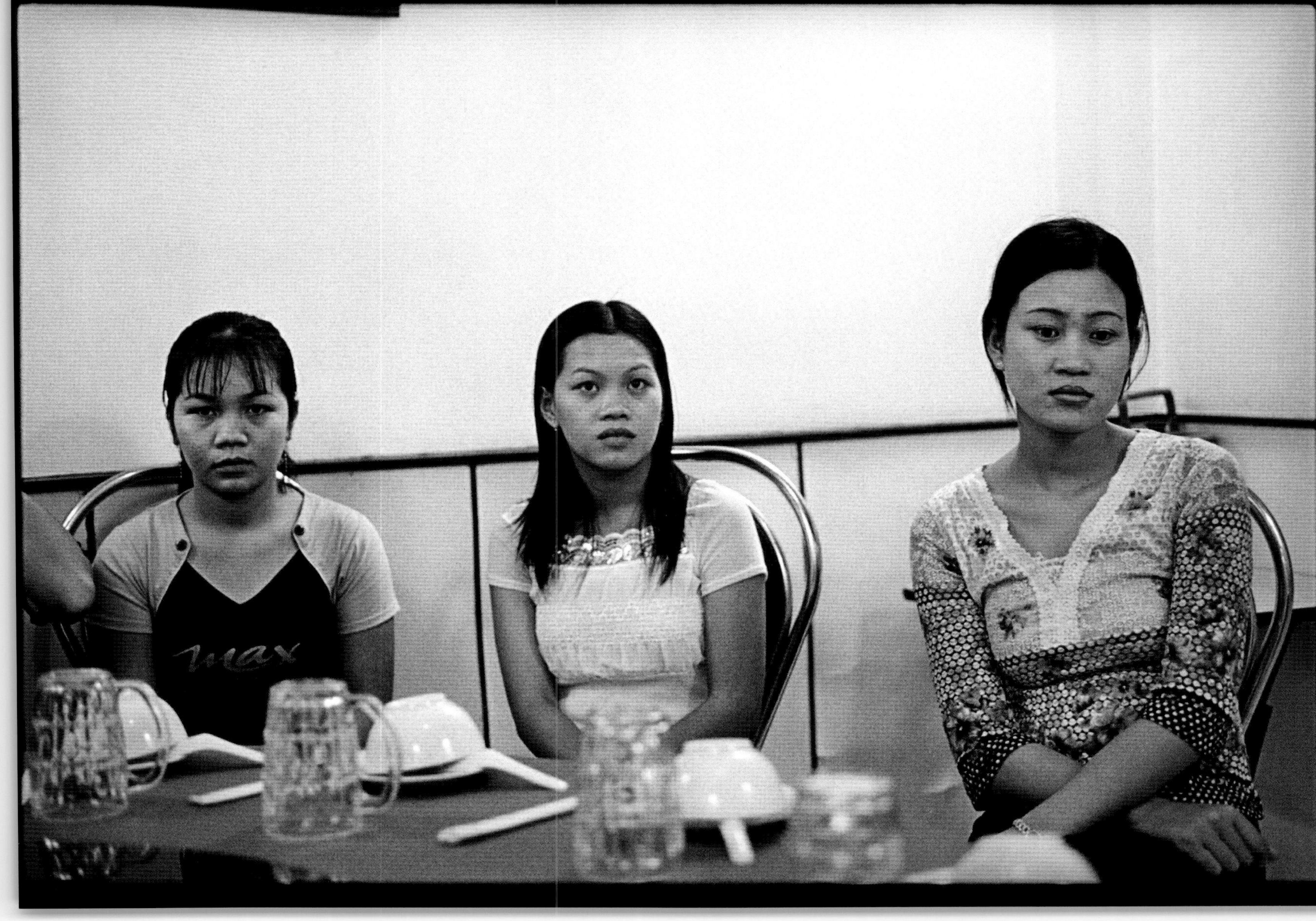
max

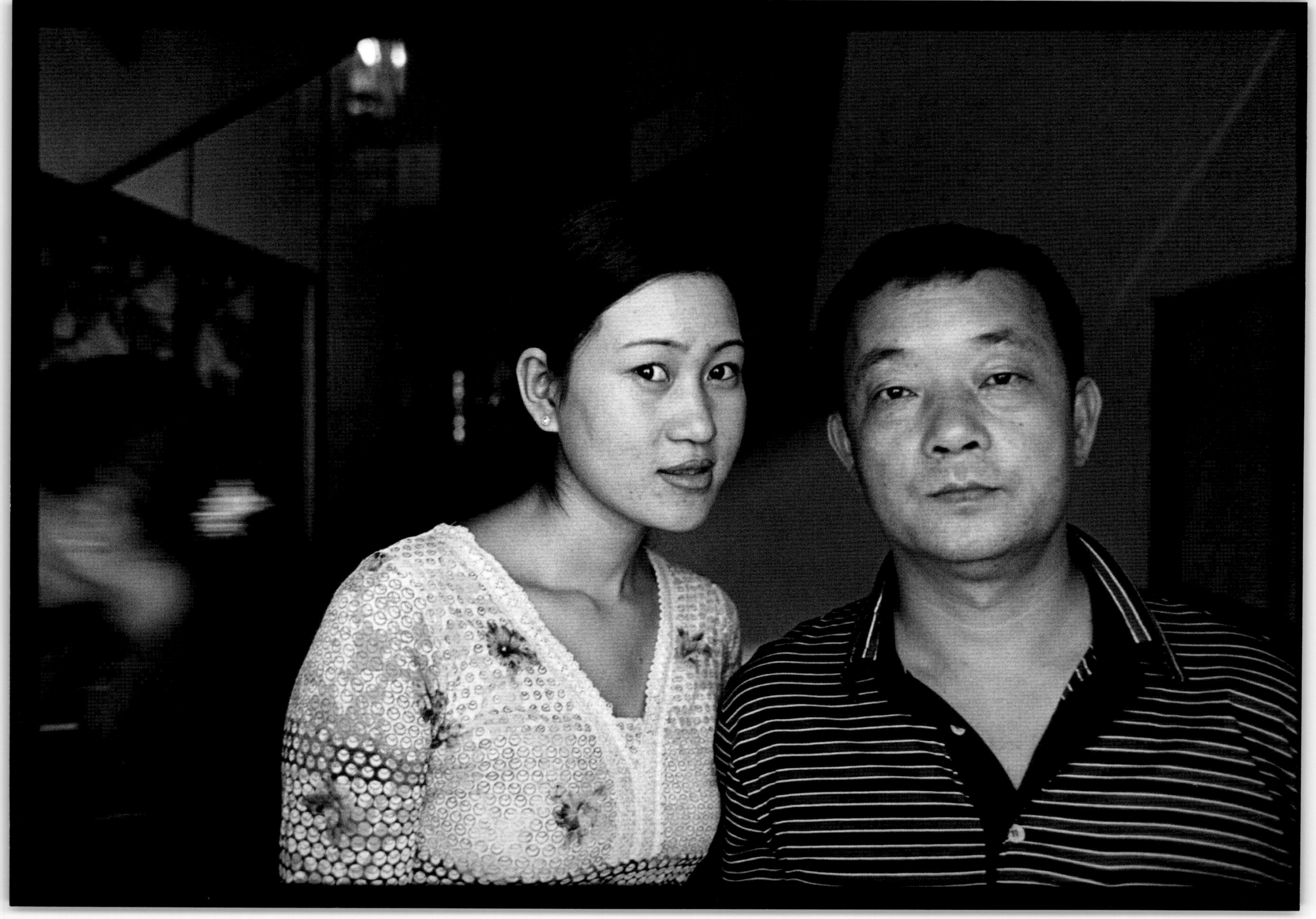

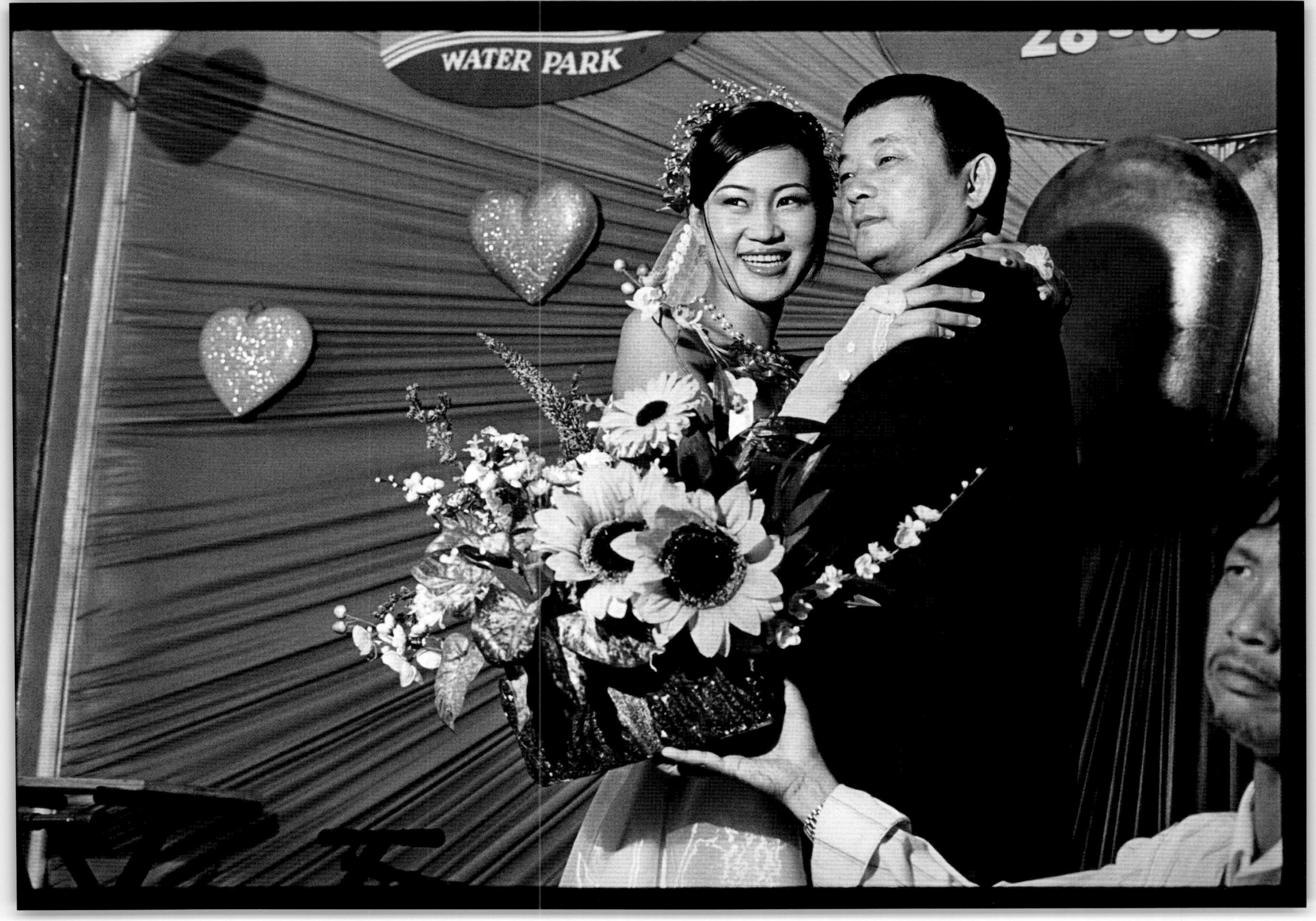
WATER PARK

On March 16, 2003, Chiang Heng-chi, 38, picked Nguyen Ngoc Phung, 23, (second from right); they were wed with three other couples on March 19.

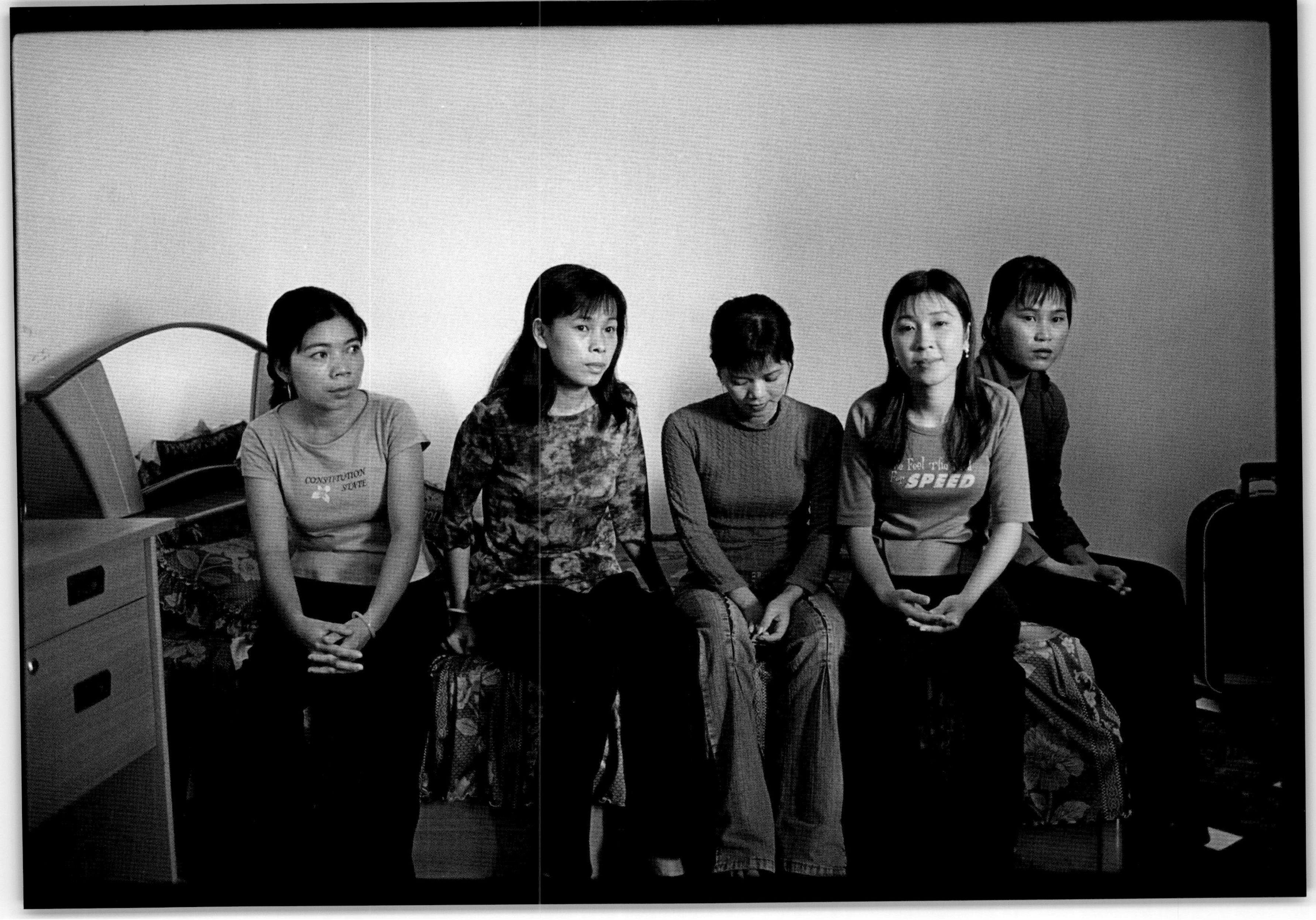
CONSTITUTION
STATE
SPEED

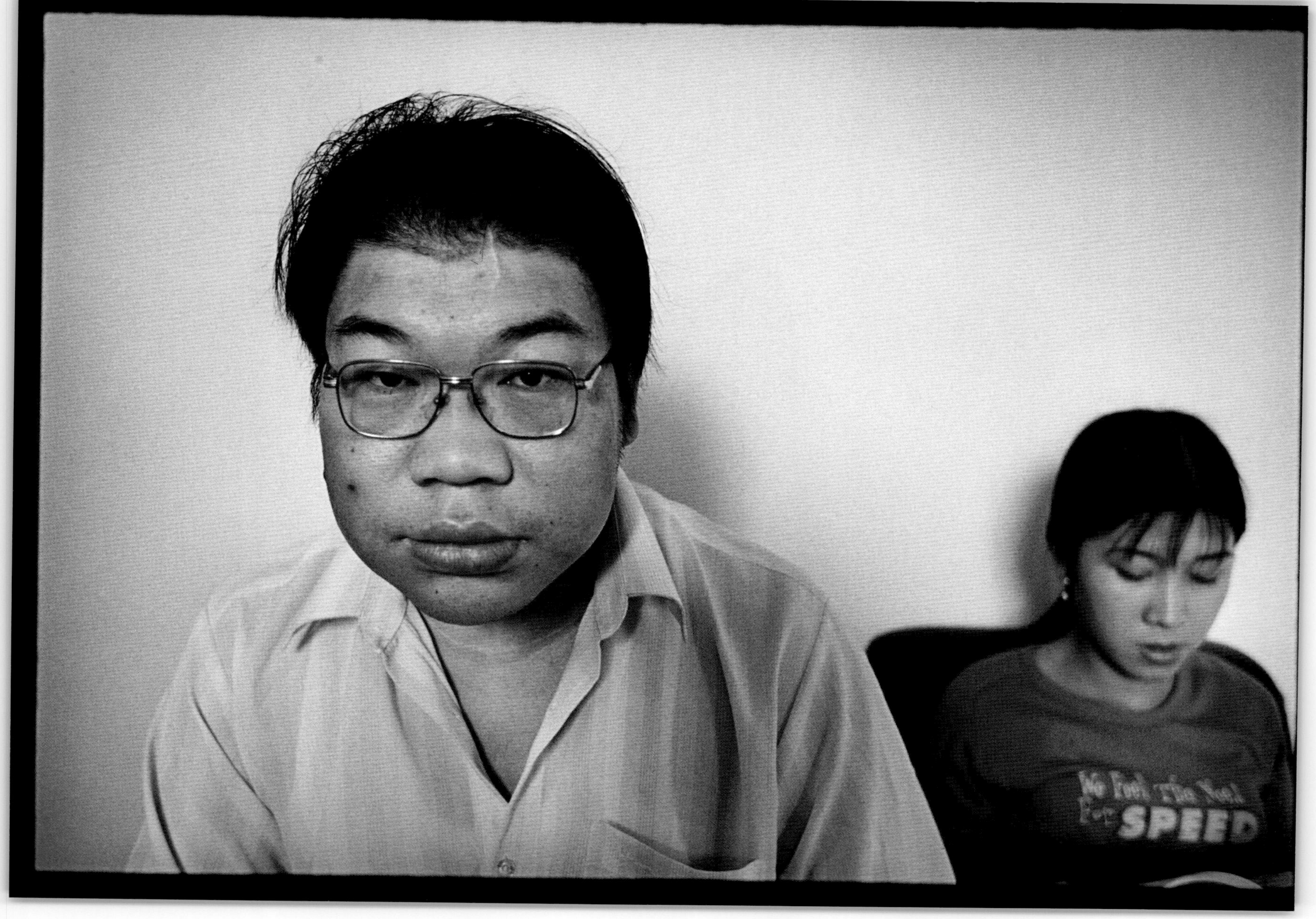
SPEED

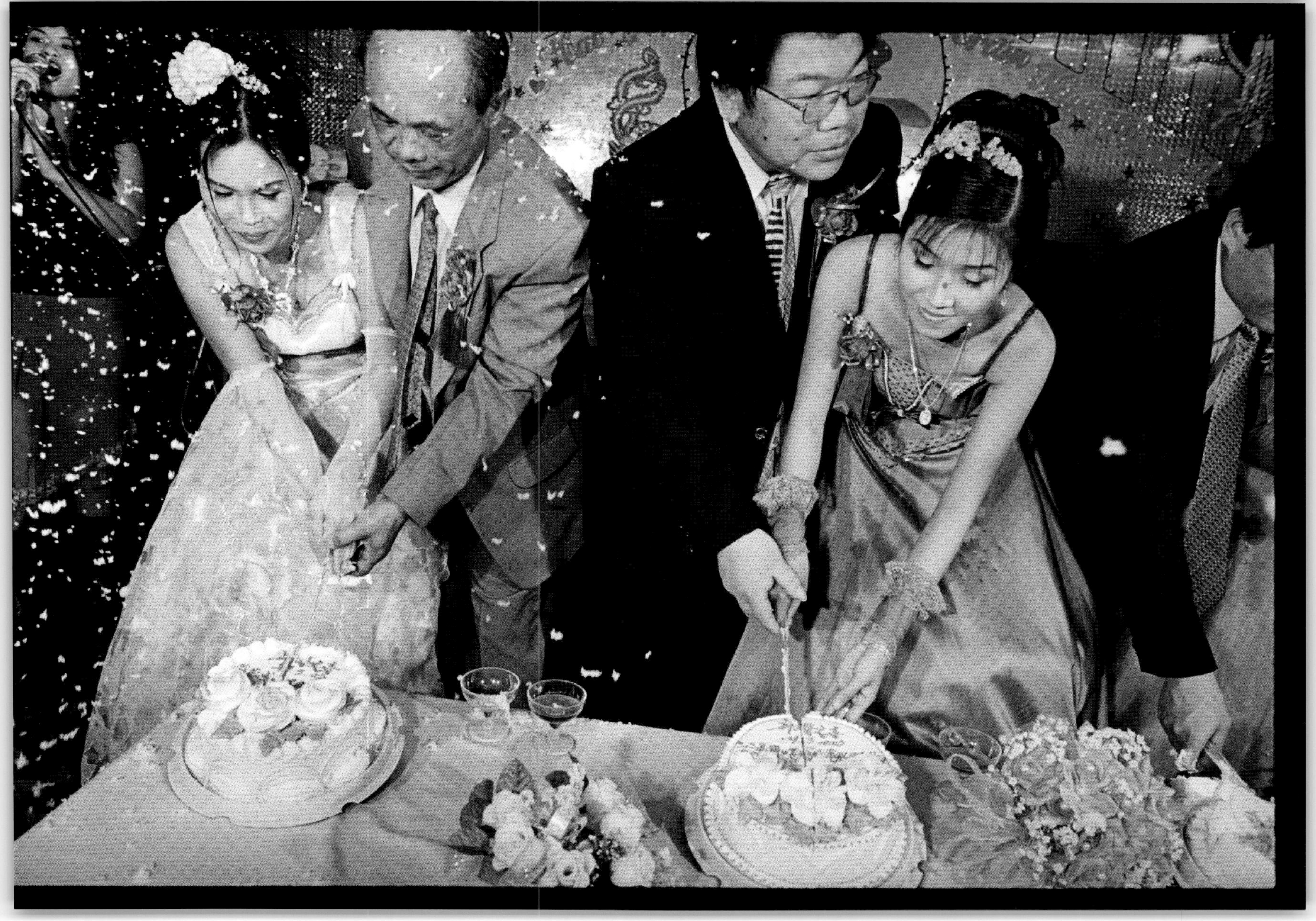

On March 16, 2003, Ngyuen Thi Cam (right) waited to be chosen; three days later she married Tsai Feng-miao.

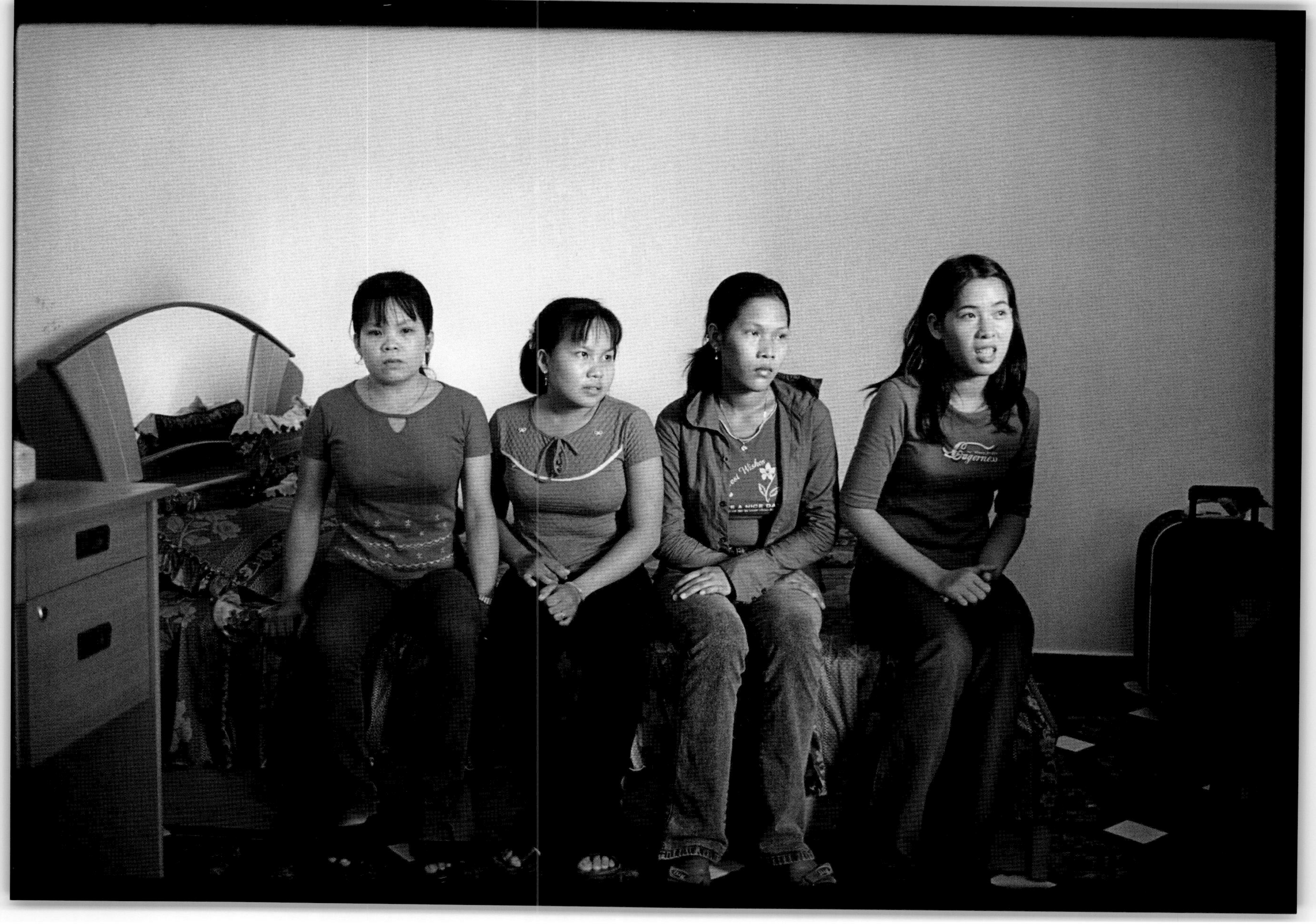

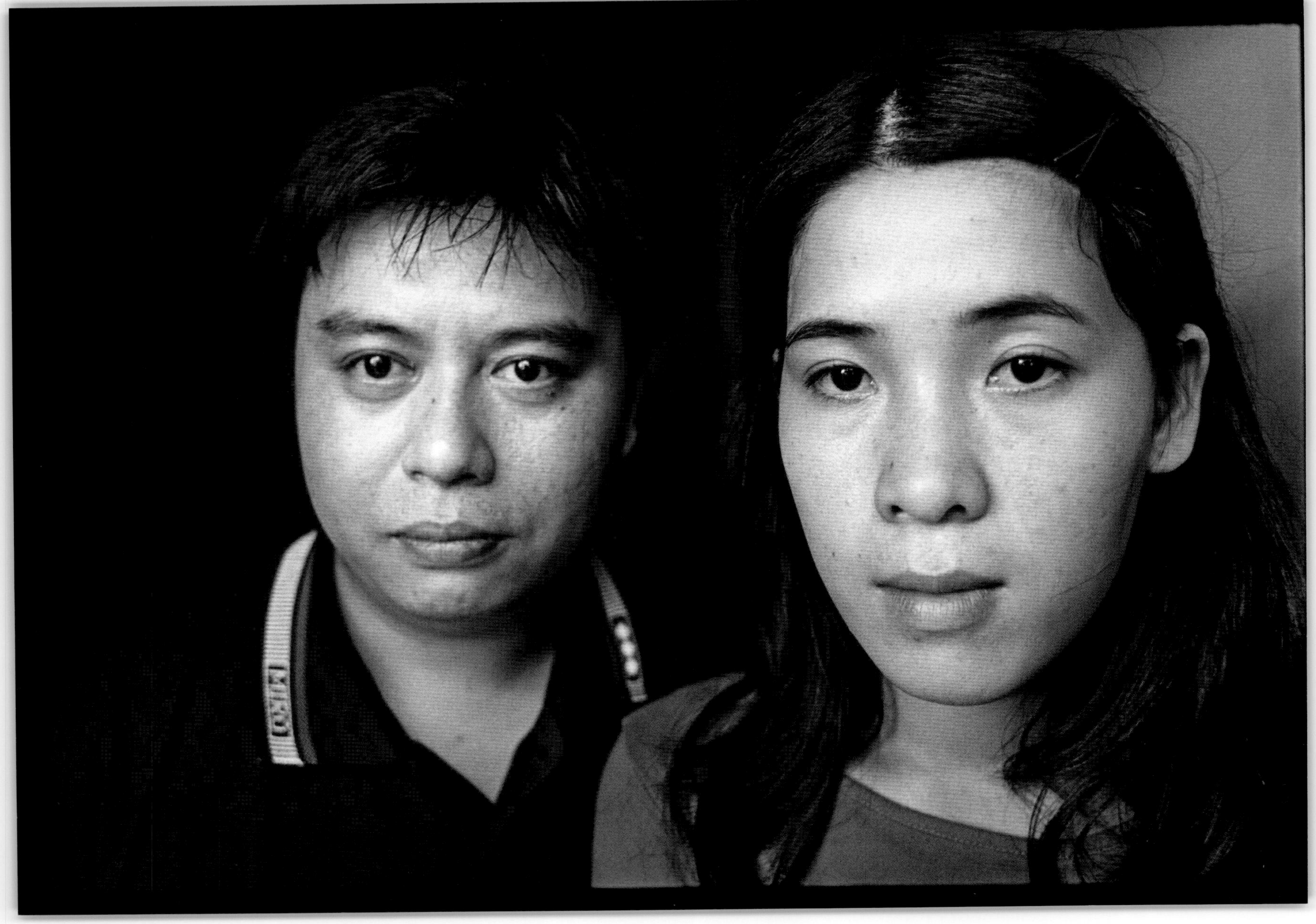

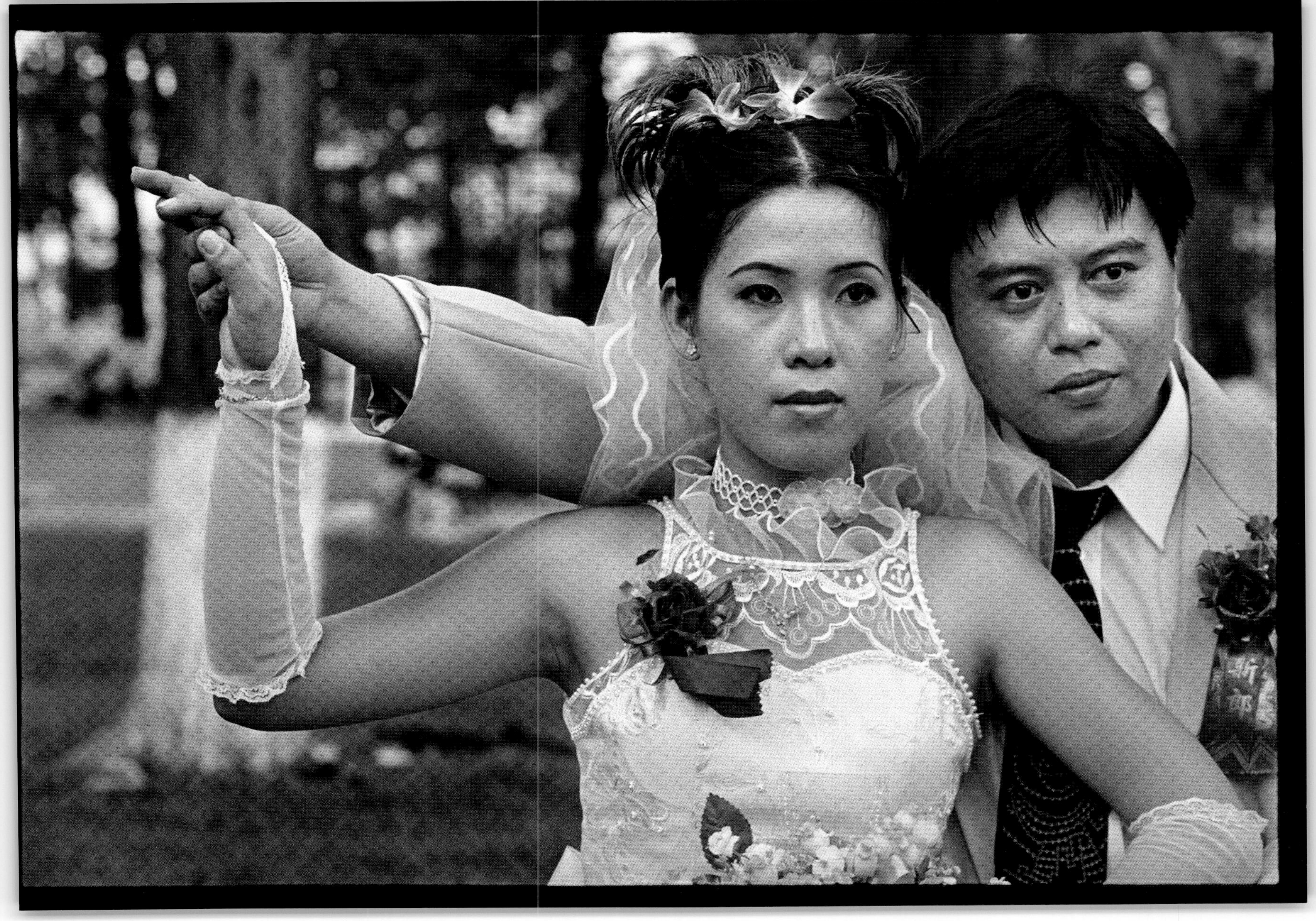
新郎

Down to the final six, Chang Wen-yeh picked Ly Thi Ngoc Hanh (center) on December 15, 2003; they were married two days later.

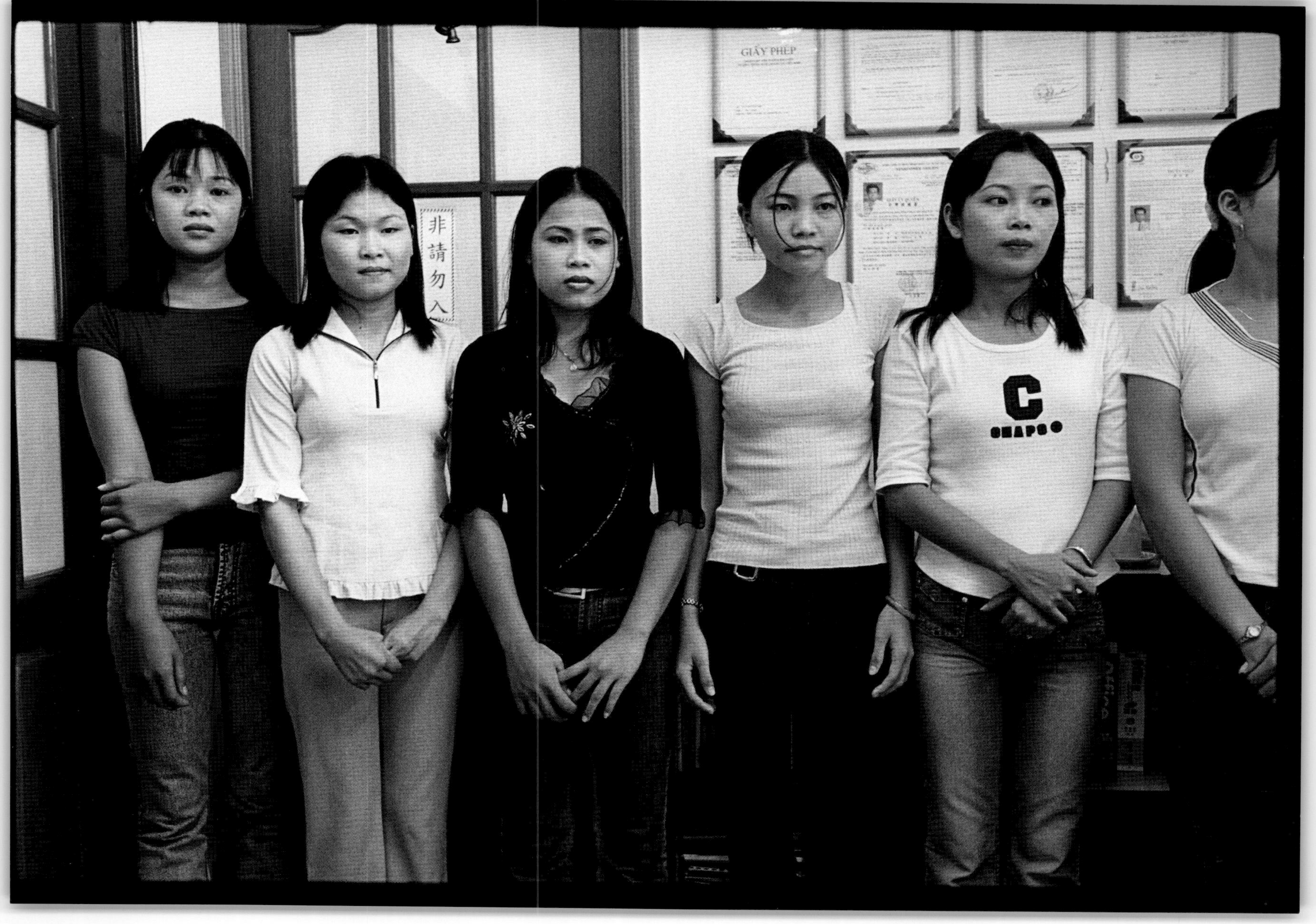
非請勿入

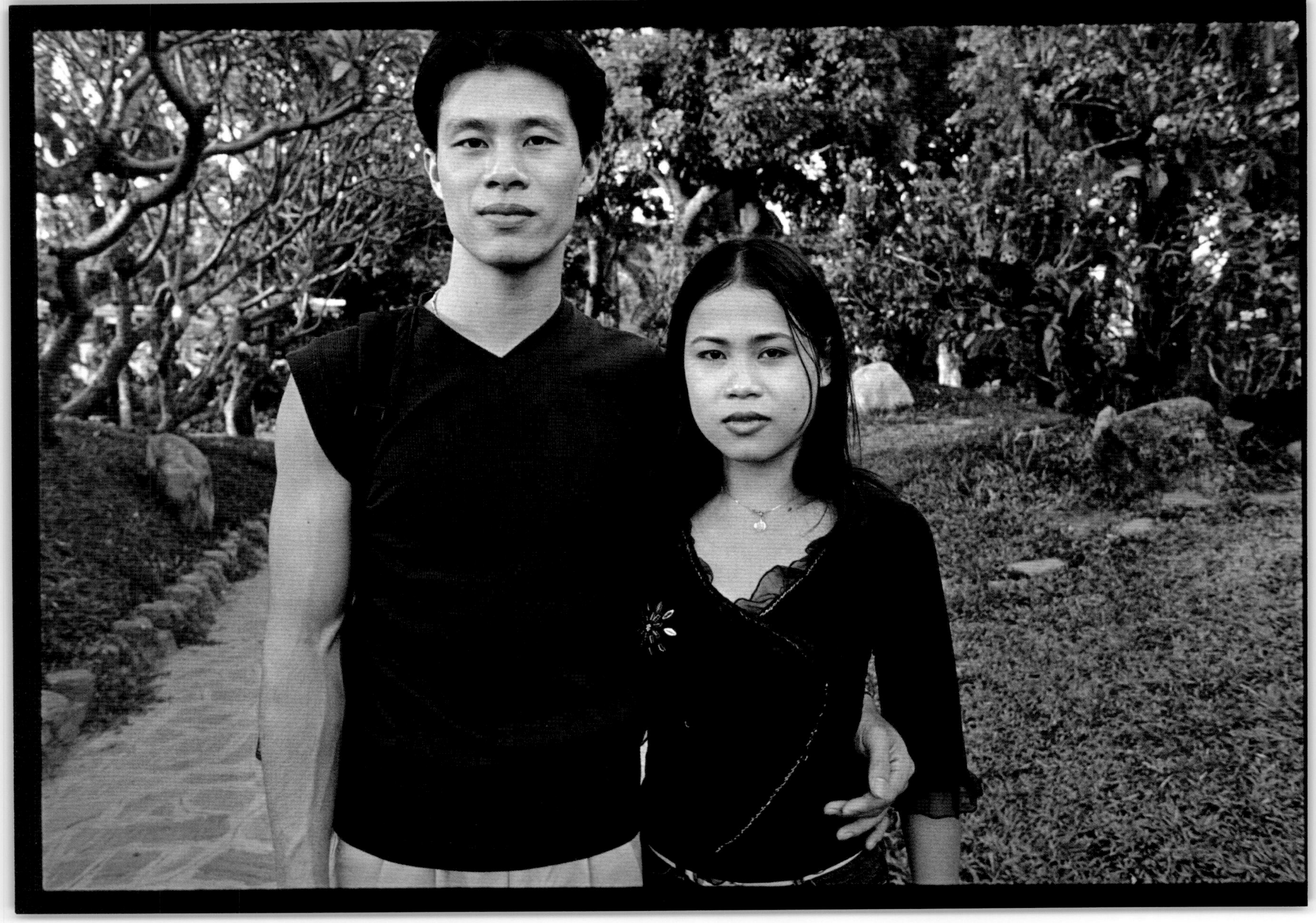

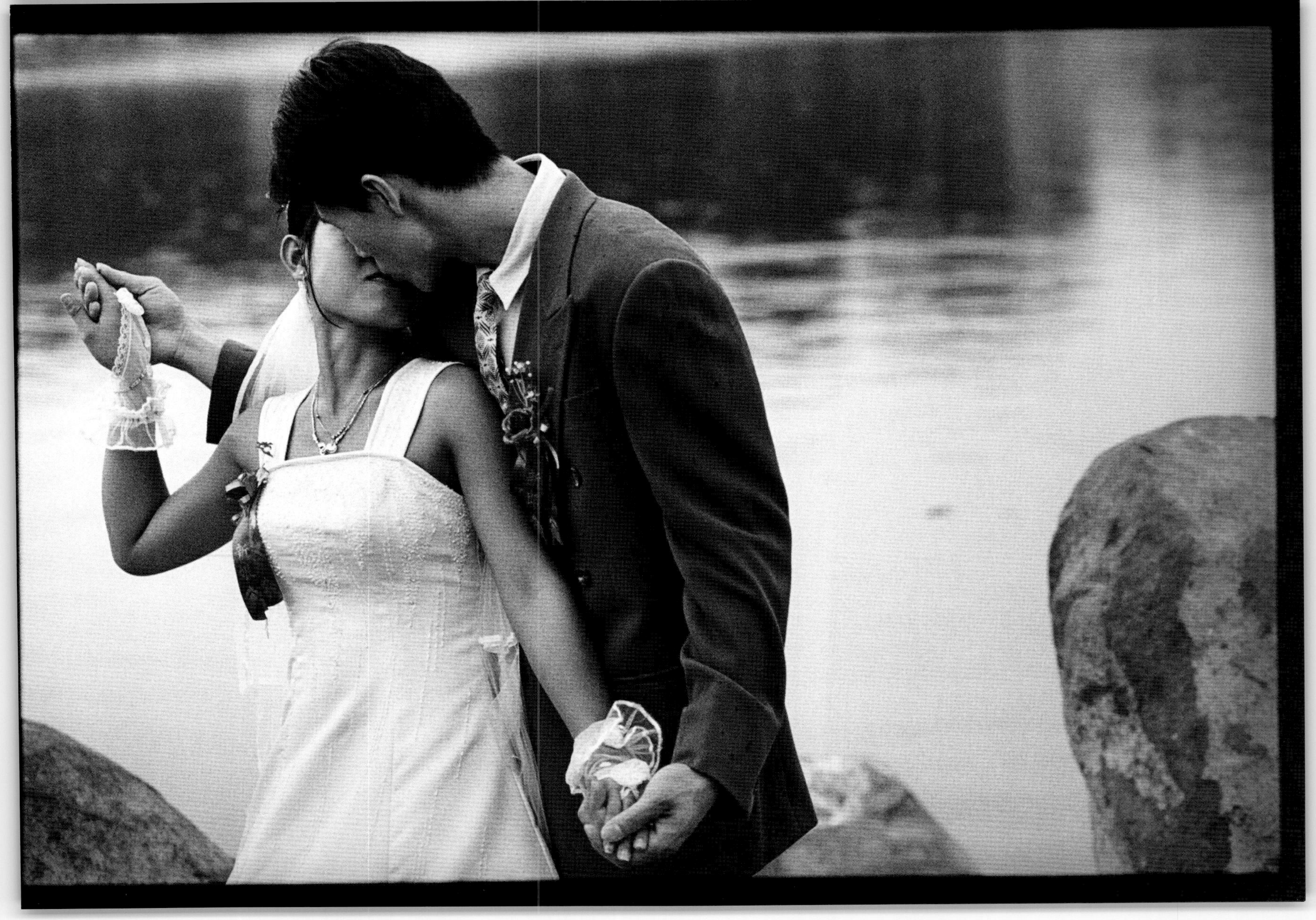

Lee Chieh-yung chose Nguyen Thi Mai (left) on July 10, 2003. They married in a double wedding on July 12.

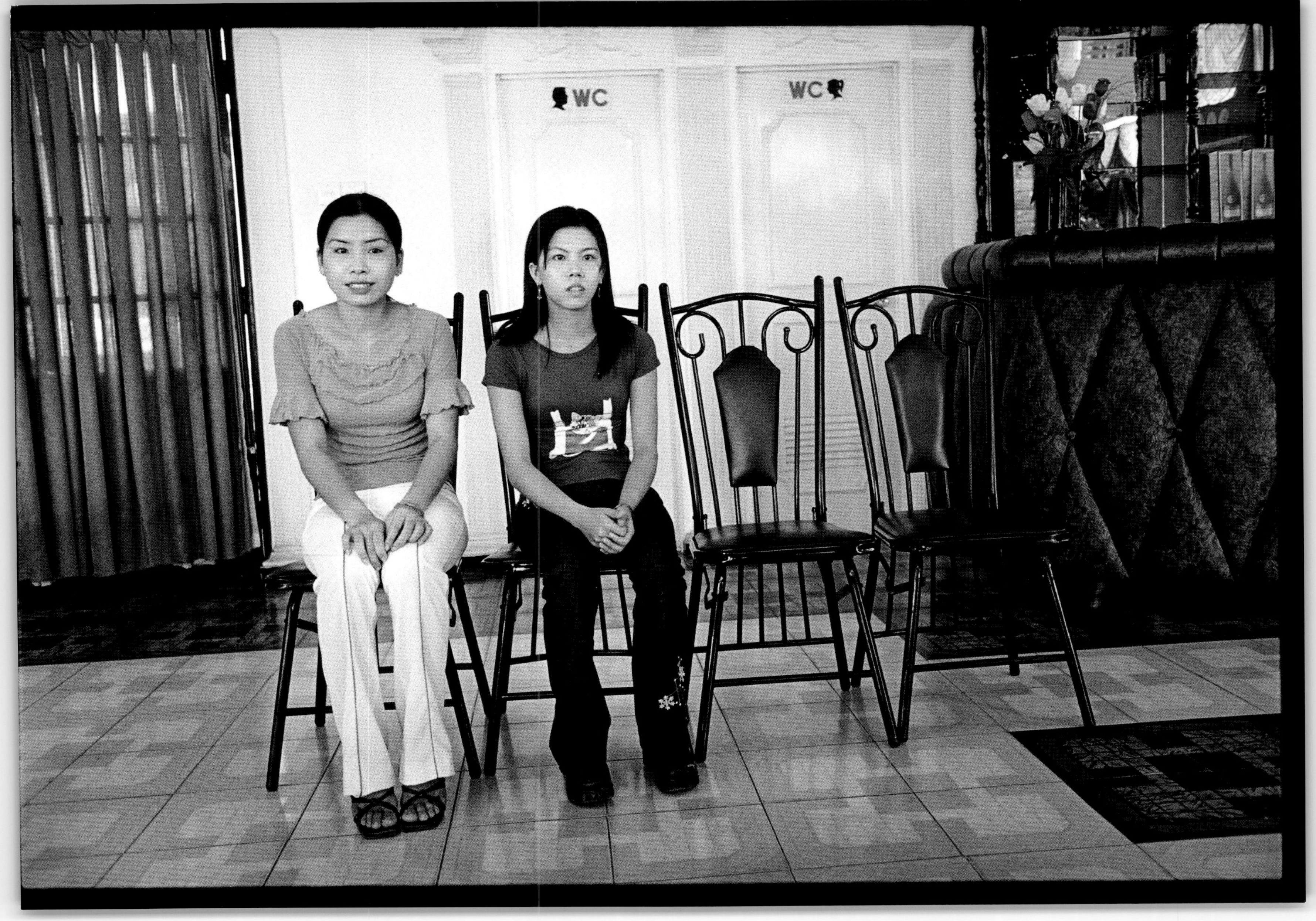
WC
WC

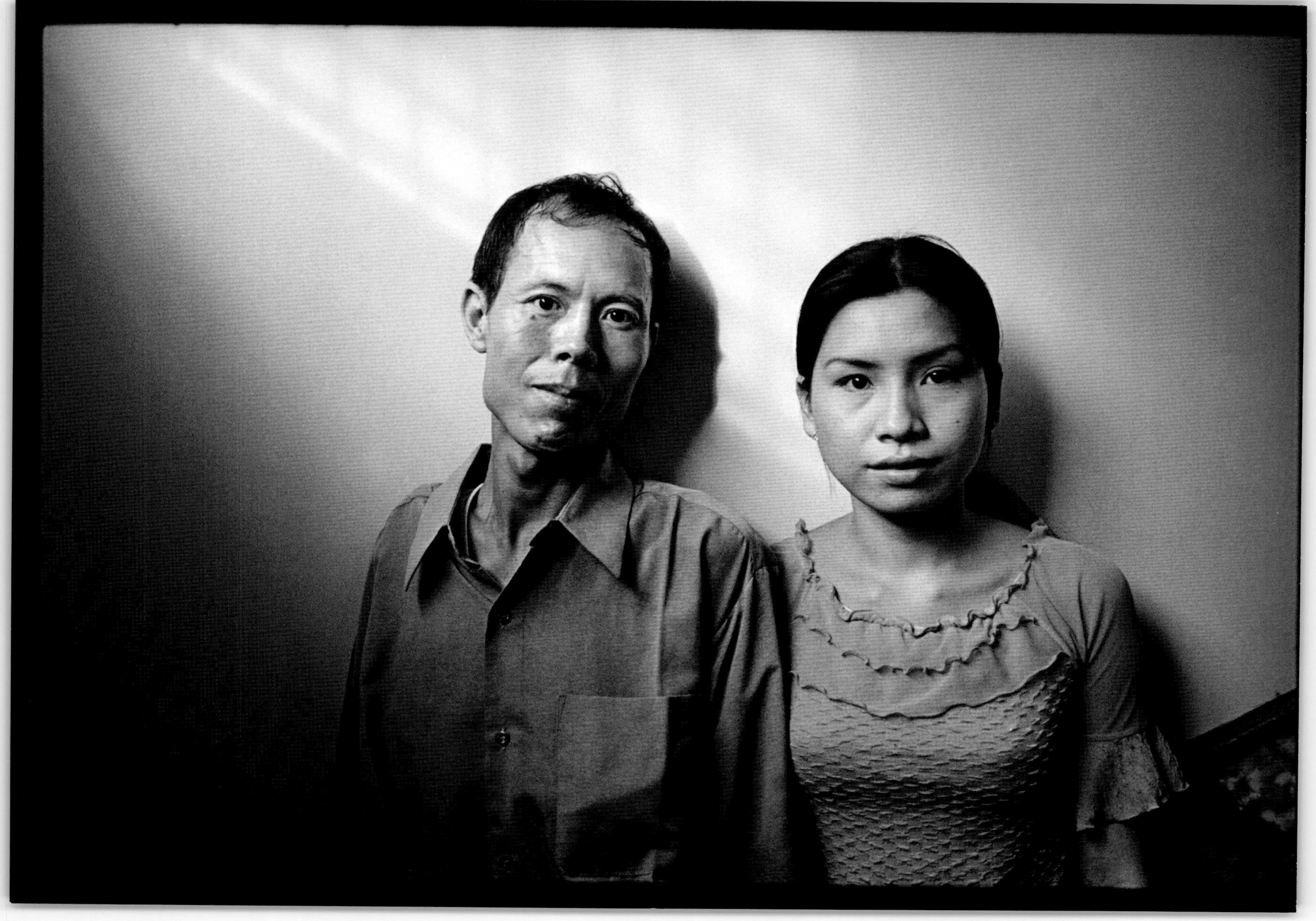

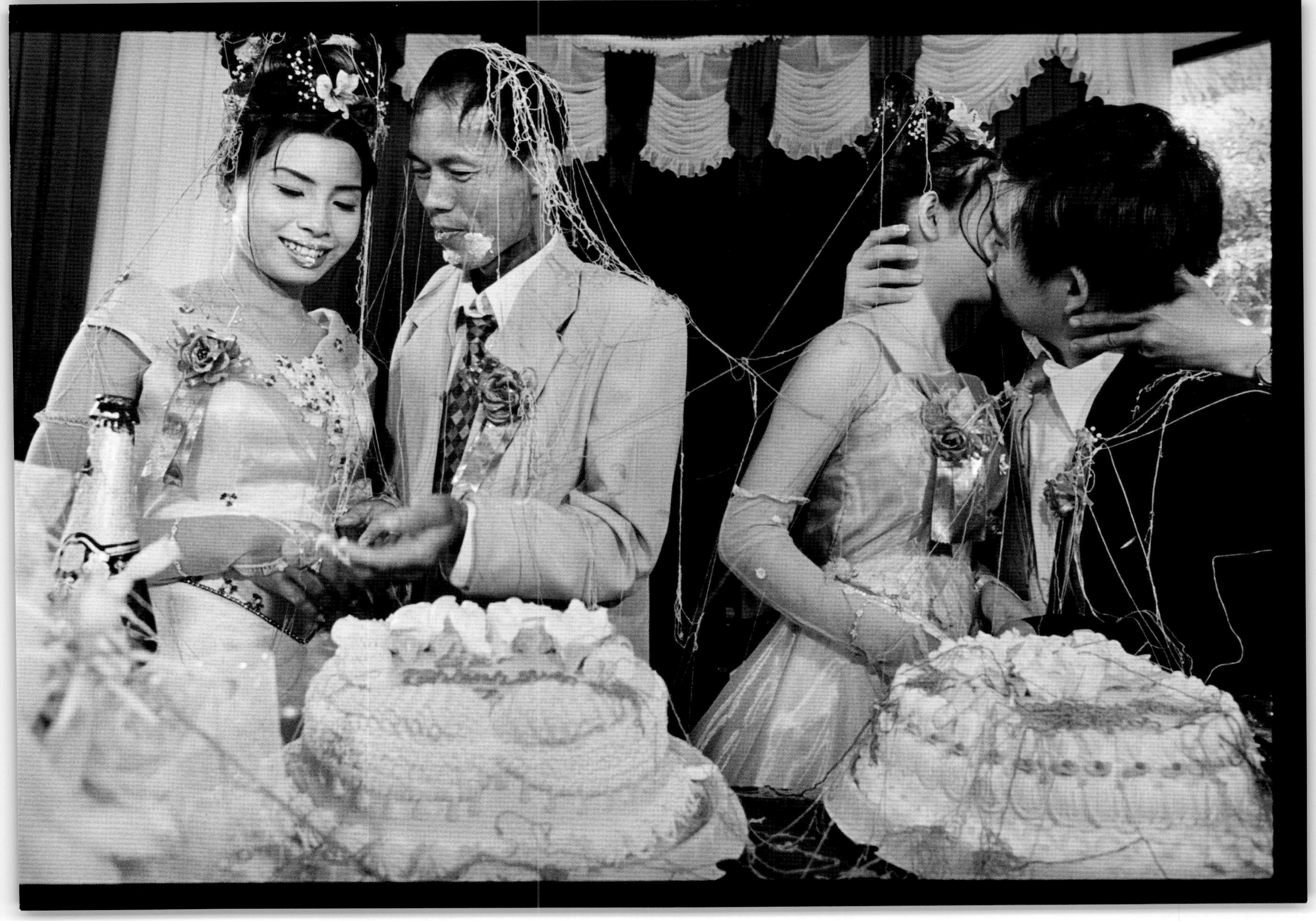

On July 10, 2003, Wang Chien-wen chose Tham Thi Bui (right); two days later they were husband and wife.

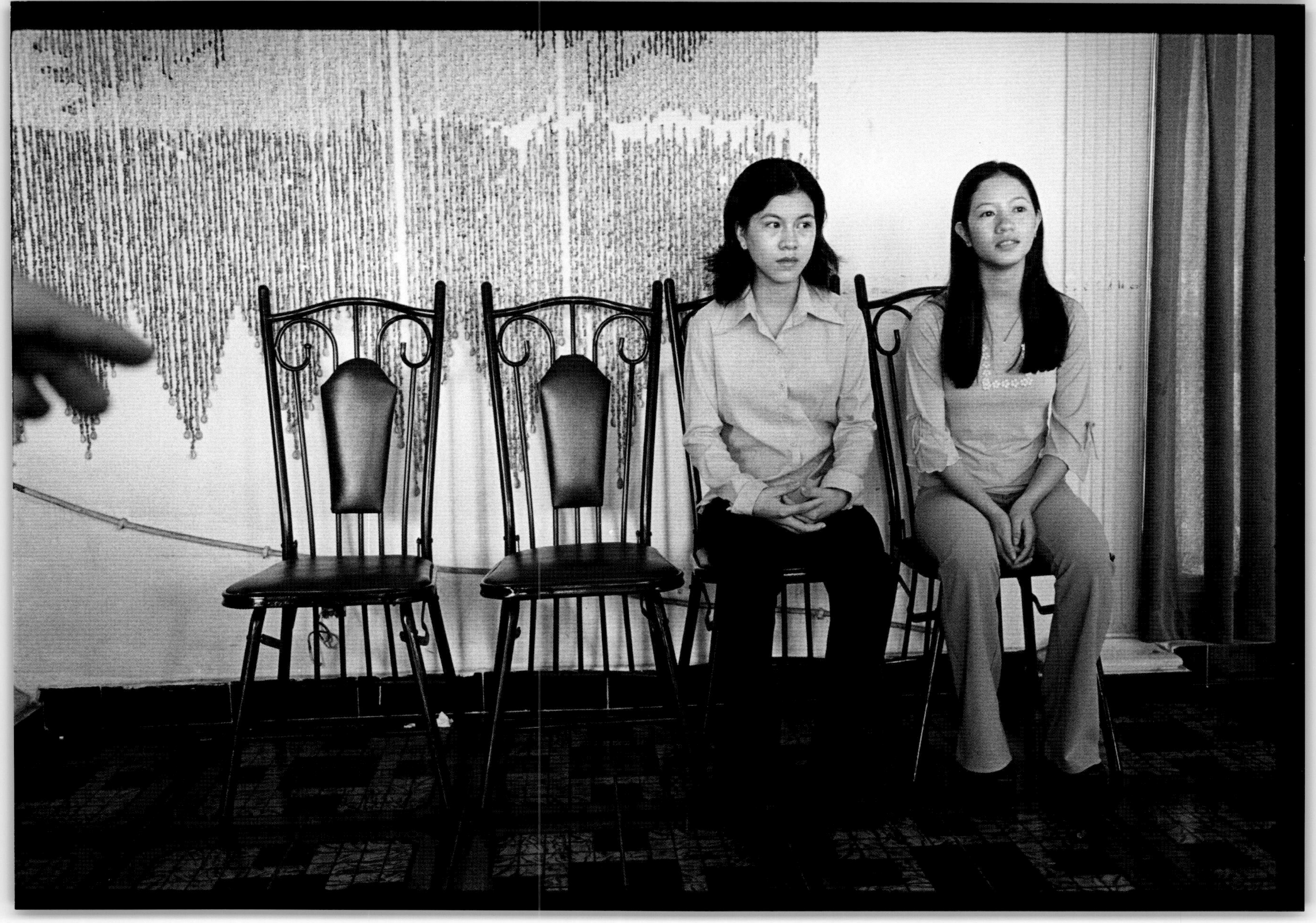

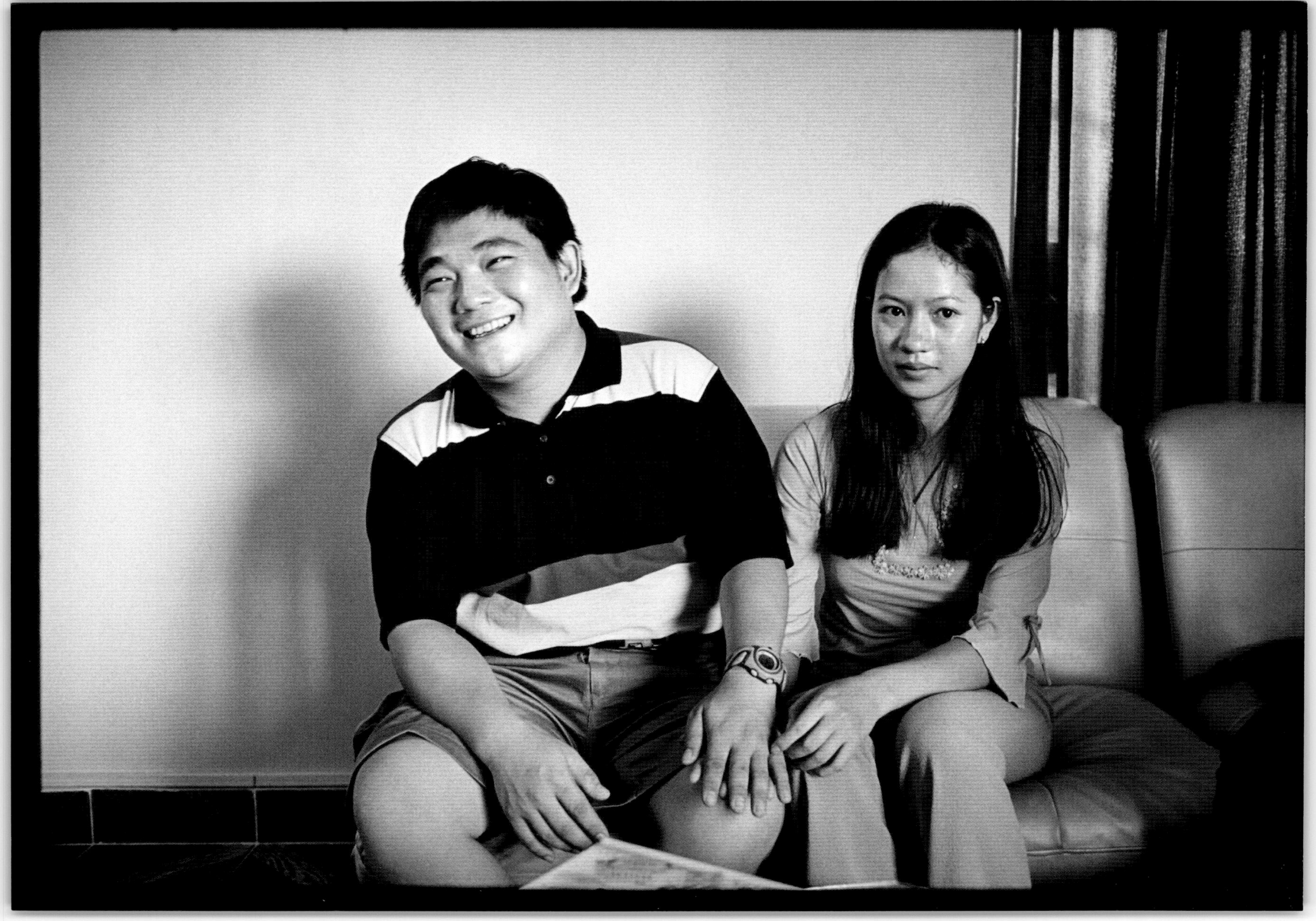

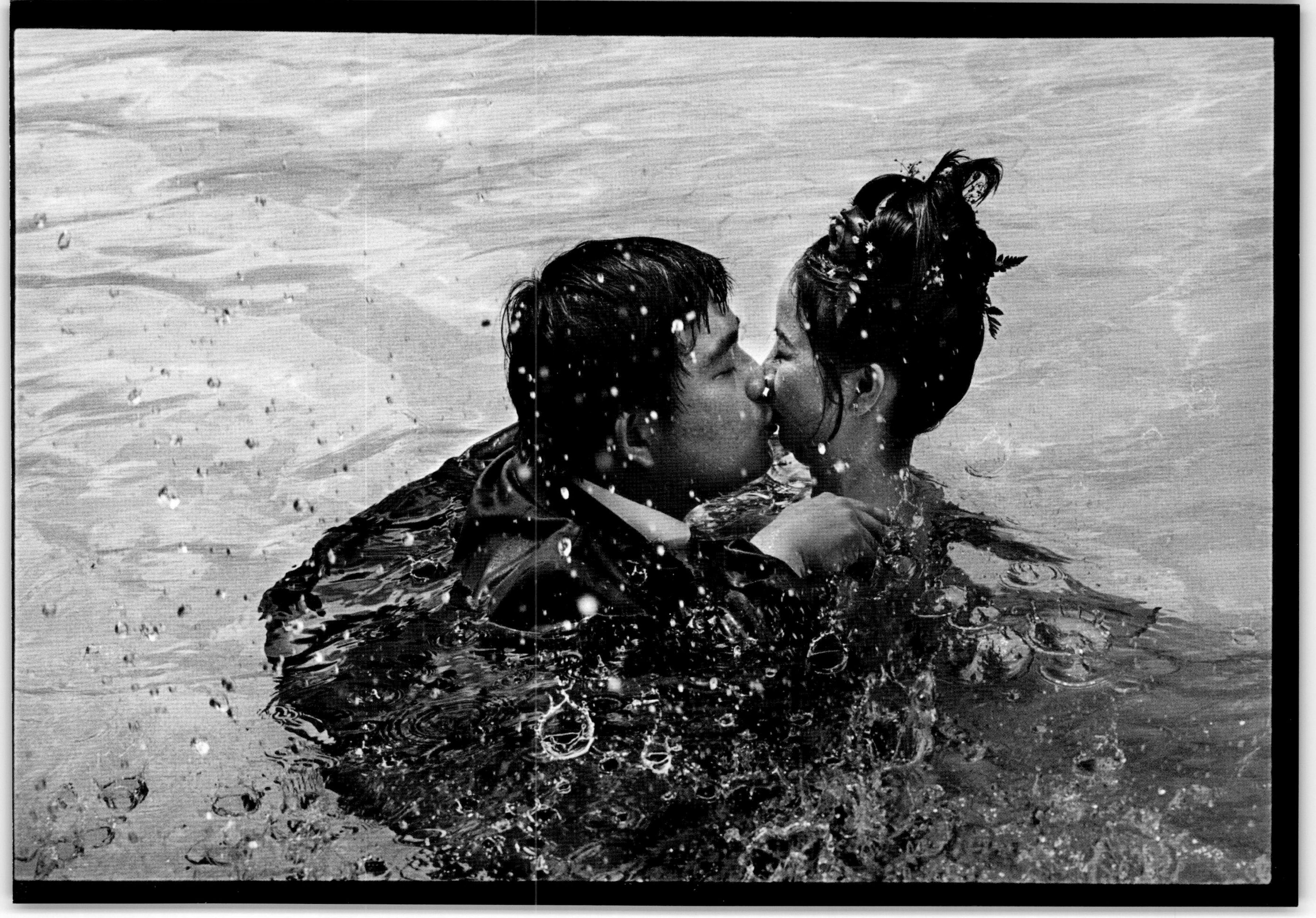

On August 24, 2003, Liu Yi-tzu had winnowed his selection down to two finalists. Two days later, he married Nguyen Thi Ngoc Chau (right).

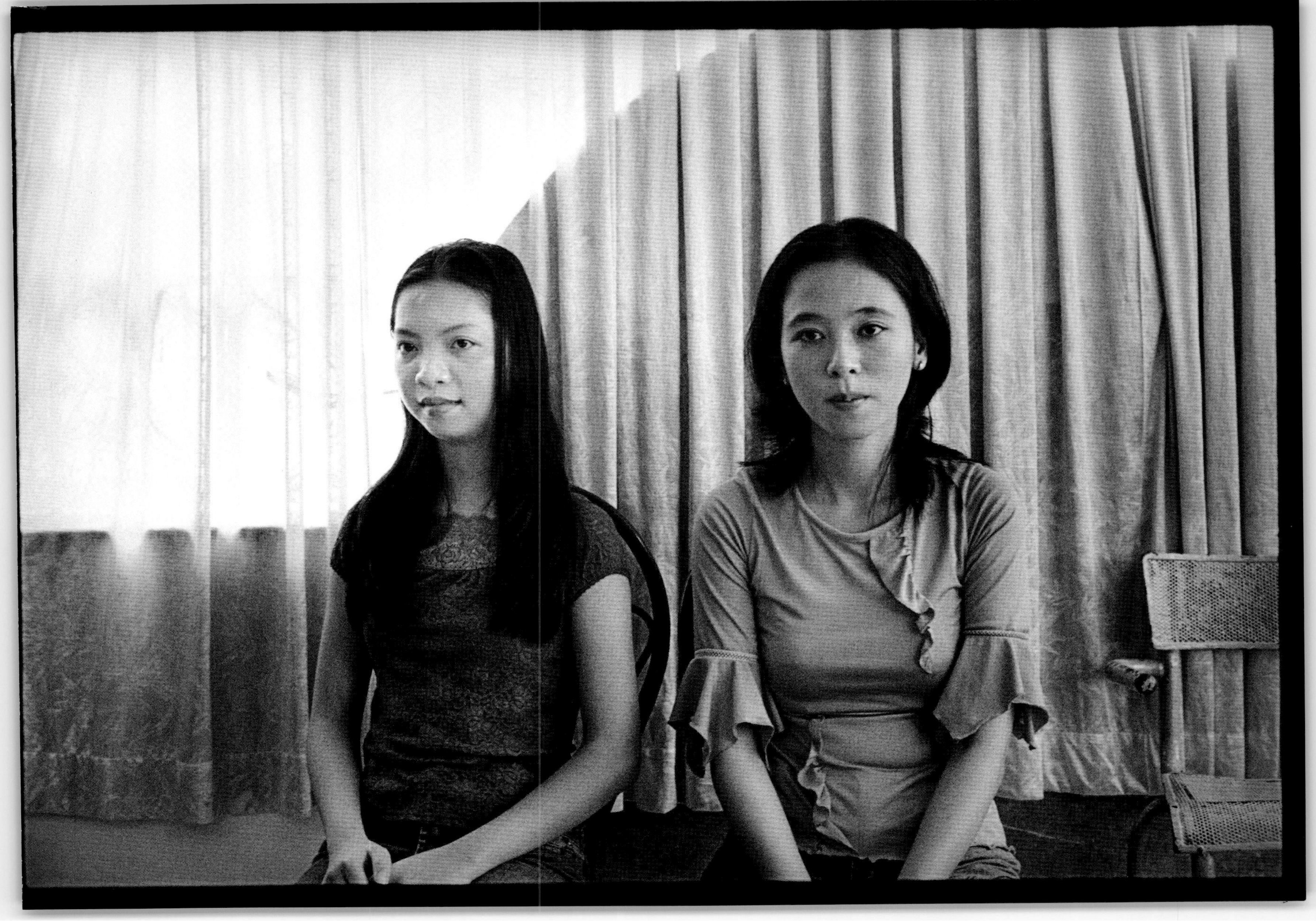

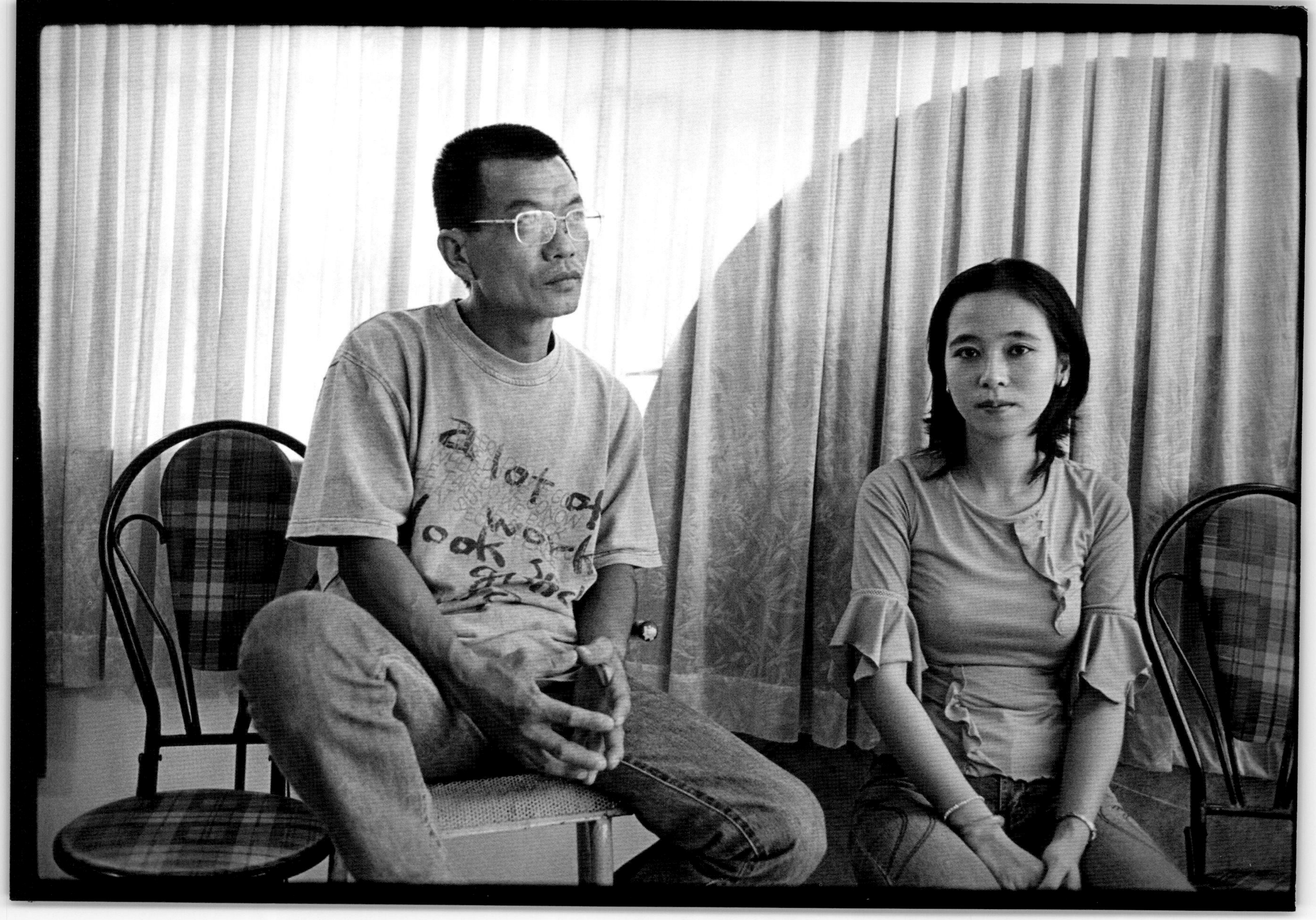
a lot of

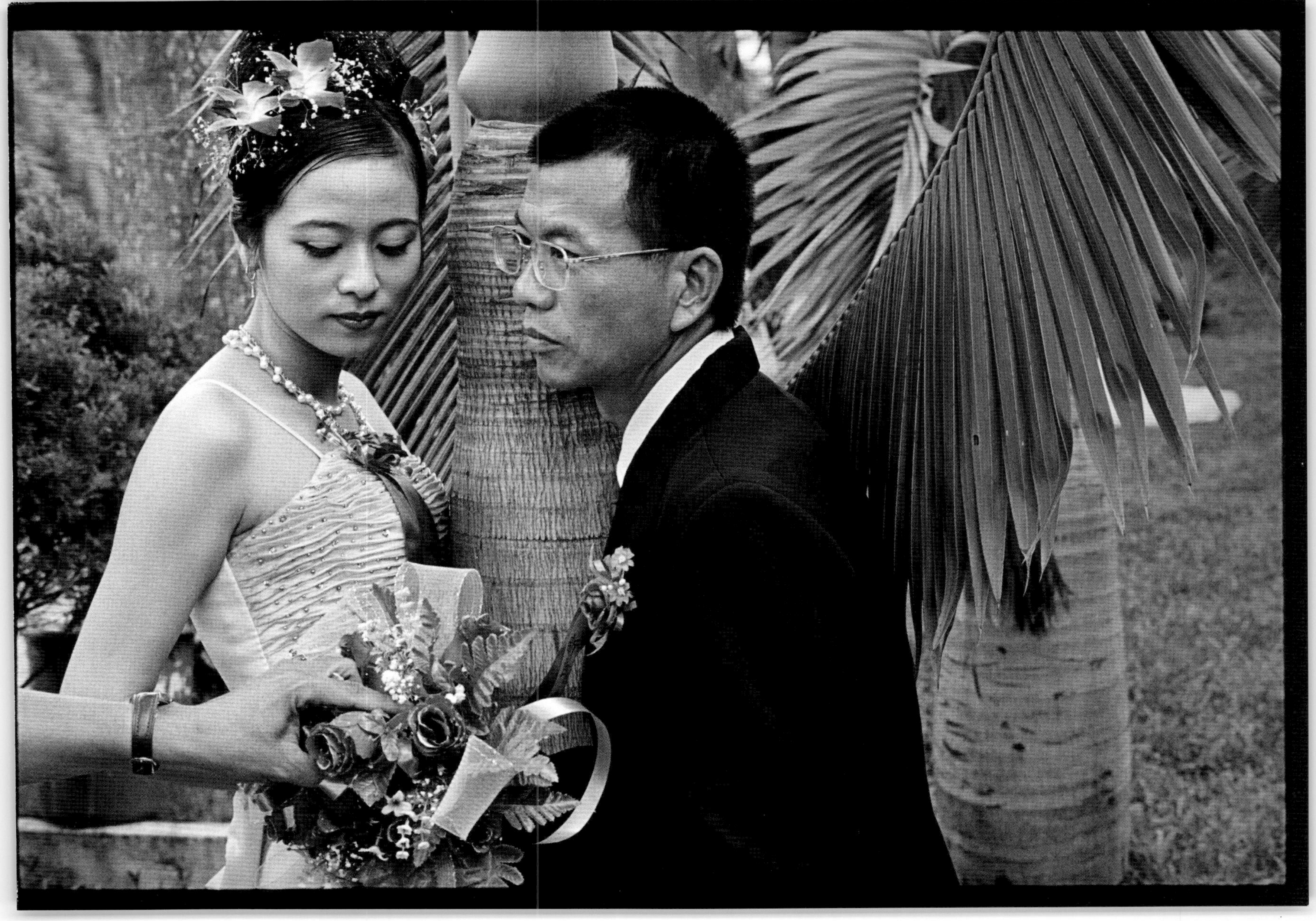

acknowledgments

I wish to thank Cheryl Lai, Chang Sheng Kai, and Linda Lee for their continued support; Lesley Martin, Francesca Richer, Diana Edkins, Bryonie Wise, Claudia Dowling, Julie Saul, James Danziger, and Mark Lubell for their insight; Alice Rose George and Howard Stein for their support of this project; and the staffs of Aperture and Magnum Photos for their dedication.

bios

Chien-Chi Chang is a member of Magnum Photos. His photographs have been exhibited at the Venice Biennial (2001), the San Paulo Biennial (2002) and the International Center of Photography Triennial (2003), among other venues. His books include *I do I do I do* (2001) and *The Chain* (2002). He lives and works in Taipei and New York.

Claudia Glenn Dowling has worked closely with photographers as a senior writer at *Life*, as writer-at-large for *People* magazine, and has been a frequent contributor to *Mother Jones* and the *Smithsonian* magazine. She now makes documentaries in New York City.

Special thanks to the Joy of Giving Something Foundation, Inc. for their support of this project.

Editor: Lesley A. Martin
Designer: Francesca Richer
Production: Bryonie Wise
Reproduction prints: Pablo Inirio

The staff for this book at Aperture Foundation includes:
Ellen S. Harris, *Executive Director;* Michael Culoso, *Director of Finance and Administration;* Lesley A. Martin, *Executive Editor, Books;* Nancy Grubb, *Executive Managing Editor, Books;* Lisa A. Farmer, *Production Director;* Andrea Smith, *Director of Publicity;* Linda Stormes, *Director of Sales & Marketing;* Diana Edkins, *Director of Special Projects;* Bronwyn Law-Viljoen, *Work scholar*

aperturefoundation
547 West 27th Street, 4th floor
New York, N.Y. 10001
www.aperture.org

Aperture is a non-profit organization devoted to photography in all its forms.

First edition
Printed and bound in Hong Kong
10 9 8 7 6 5 4 3 2 1

Library of Congress Control Number: 2004098599
ISBN 1-931788-56-1

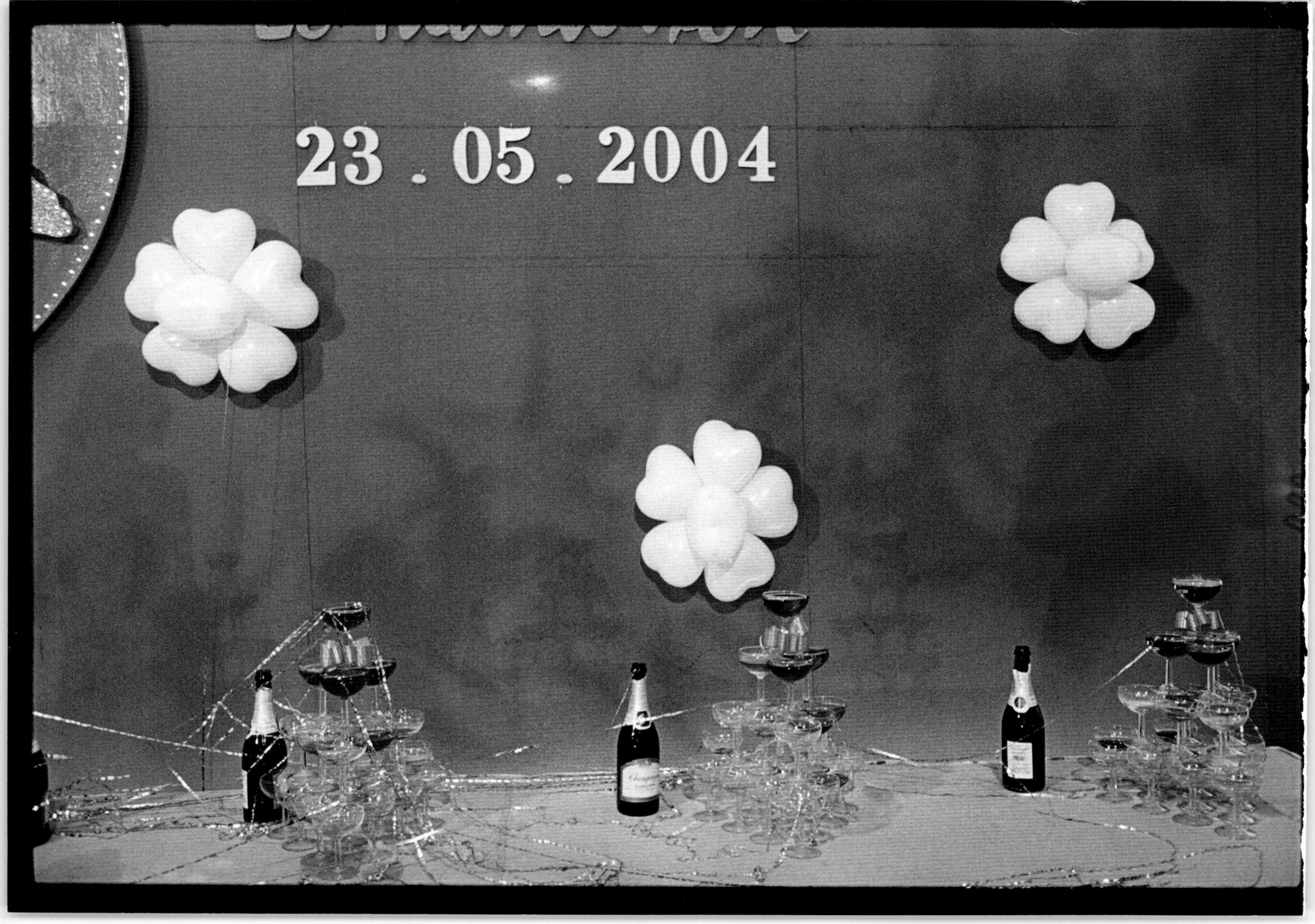
23 . 05 . 2004